Behavioral Systems and Nursing

Jeanine Roose Auger
Nursing Service
UCLA Neuropsychiatric Institute

prentice-hall, inc.
englewood cliffs, n.j.

Library of Congress Cataloging in Publication Data

Auger, Jeanine Roose.
  Behavioral systems and nursing.

  (Prentice-Hall scientific foundations of nursing
practice series)
  Bibliography: p. 197
  Includes index.
  1. Nursing—Psychological aspects. 2. Human behavior. I. Title. [DNLM: 1. Behavior. 2. Systems
analysis. 3. Nursing. WY87 A919b]
  RT86.A88   610.73'01'9   75-34299
  ISBN 0-13-074484-0

prentice-hall
scientific foundations of nursing practice series

Dorothy E. Johnson, Editor

Cover photographs by Englewood Hospital, Englewood,
N.J.; Harold M. Lambert from Frederic Lewis, Inc.;
and Elizabeth Wilcox.

10 9 8 7 6 5 4 3 2 1

Printed in the United States of America

PRENTICE-HALL INTERNATIONAL, INC., London
PRENTICE-HALL OF AUSTRALIA PTY. LIMITED, Sydney
PRENTICE-HALL OF CANADA, LTD., Toronto
PRENTICE-HALL OF INDIA PRIVATE LIMITED, New Delhi
PRENTICE-HALL OF JAPAN, INC., Tokyo
PRENTICE-HALL OF SOUTHEAST ASIA PTE. LTD., Singapore

Contents

Foreword

The growth of nursing as a profession and a scientific discipline has been rapid in recent years. New orientations and emphases in nursing education are readily apparent. New organizational patterns to facilitate professional practice are being sought and found. Nursing practice becomes increasingly significant and sophisticated as knowledge is developed along many fronts and is brought to the service of patients. In the face of these changes, and because the knowledge potentially available for practice no longer can be contained within the boundaries of the comprehensive course textbook approach in education, this new approach to the subject matter underlying nursing practice is offered.

    The PRENTICE-HALL SCIENTIFIC FOUNDATIONS OF NURSING PRACTICE SERIES consists of a number of relatively brief monographs, each dealing with a specific and circumscribed topic. Each monograph is designed to order and integrate pertinent knowledge drawn from widely dispersed sources. Collectively, the series will attempt in time to cover all essential areas, thus providing the fullest scope possible. Such an approach appears to have a number of advantages. The treatment of single topics in separate books permits great flexibility in use in educational programs that vary markedly in the structure and level of courses offered and in the content encompassed or emphasized. It allows the examination of each topic in greater depth than is possible when many issues or concepts must be considered in a comprehensive text, and it encourages consideration of new and emerging foci of interest and development. It avoids the sometimes disjointed nature of publications with multiple authors while simultaneously offering presentation of diverse topics by those especially well qualified to do so. And, finally, it may encourage the development of specialized interest and competence by the advanced practitioner.

The major emphasis in this series is on selected nursing problems. The first concern is for problems that patients are likely to experience in any setting in which nursing is practiced, even though the patients will differ along such dimensions as age; sex; medical diagnosis; and degree, severity, and chronicity of the illness. To the degree made possible by existing knowledge, these problems are described and explained, and the scientific basis for their identification and nursing management is provided. The tentative nature of the basis for prescription in particular will be emphasized, and it is hoped that the presentation will excite in the student or practitioner the skepticism and challenge essential to future contributions to knowledge in the field.

Other volumes in the series cover other aspects of nursing as a field of study and practice. One provides an introduction to the profession and the discipline. Others attempt to synthesize and order knowledge about human behavior in health and illness as a basis for understanding the nature of nursing problems. Still others examine the diagnostic process in nursing, modes of scientific inquiry, and certain treatment methods. Both the underdeveloped nature of nursing knowledge and the steady growth of that knowledge demand that the series remain open-ended, flexible, and subject to early revision.

The scope of nursing practice is broad, and a wide range of knowledge and skills is required among its practitioners. No series of monographs or group of texts can provide adequately for the needs of all students or practitioners in the field. This series is focused upon the need of baccalaureate and higher degree students but will also serve the needs of many others. The PRENTICE-HALL SCIENTIFIC FOUNDATIONS OF NURSING PRACTICE SERIES is an attempt to place in the hands of this group stimulating material of outstanding quality in subject matter and presentation.

Dorothy E. Johnson
Professor of Nursing
University of California
Los Angeles

Preface

The practice of nursing and the education of new practitioners has been profoundly influenced in the past three decades by the explosion of knowledge in the behavioral and natural sciences. Basic nursing curricula routinely include courses in the areas of psychology, anthropology, sociology, and education, in addition to the traditional courses of anatomy, physiology, bacteriology, physics, and chemistry. Unfortunately, the student is often confronted with a fragmented curriculum in which the immediate relevance and application of the content learned in these related fields is unclear or unavailable. Either the students must assume the responsibility for formulating their own framework of relevancy, or the concepts remain fragmented and compartmentalized in their approach to the care of patients.

The value of a model of nursing is that it can organize and integrate these diverse bits of knowledge into a comprehensive framework that can then be applied by student and practitioner to the care of patients. The purpose of this text is to present one model of nursing, the behavioral systems approach, developed by Dorothy Johnson, R.N., M.P.H., Professor of Nursing at the University of California, Los Angeles. Ms. Johnson has been, and continues to be, a major force in the development of a scientific foundation to nursing practice. She has devoted her professional and academic career to the science of nursing, which reflects her belief that nursing shares with all professions an obligation that goes "beyond accepting the current state of affairs to shaping the reality of the future."[1]

This book includes the presentation of general systems theory followed by the specific application of the theory to nursing as formulated by Ms. Johnson. Behavior is viewed as a function of multiple interrelated factors including bio-psycho-social-cultural factors, the age, sex, and learning history of the individual and the immediate environmental situation in which the behavior is observed. These factors are discussed in relation to each of the eight subsystems of behavior: ingestion, elimination, sexual, affiliation, protection-aggression, dependency, achievement, and restoration. The assessment and analysis of behavior using a systems model framework is presented; the emphasis of the assessment is upon relating "current" behavior to "usual patterns of behavior" for the individual. Finally, two case studies are presented in order to demonstrate the direct application of the approach to two clinical settings: a neurology medical unit and a psychiatric outpatient department.

Based upon a knowledge of the behavioral systems approach, the nurse is able to evaluate the status of the person who is ill and the significance of changes that may or may not have occurred in patterns of behavior. This content will help the student to develop a broader concept of the relationship between health and illness, the wide variations of "normal" behavior, and changes that may occur as a consequence of illness and/or hospitalization. It will also provide a means for developing scientific, rational interventions designed to attain the goal of supporting, maintaining, and protecting the individual during the process of recovery.

The concepts that have been included in the body of the text are by no means exhaustive of the current knowledge that is useful and pertinent to the understanding of behavior. An attempt has been made to select a broad base of concepts that are representative of the individual academic disciplines and to demonstrate their potential integration through the application of the principles of behavioral systems theory. It is assumed that the student will have completed the basic courses in each of the disciplines before studying this model. As knowledge expands and becomes more precise, classroom examples and lectures can incorporate the new concepts into the already existing framework. It is recommended that the student and practitioner consult the current professional journals for literature within each of the disciplines that are related to the study of human behavior. One of the strengths of this model is the flexibility with which new findings can be related to already existing knowledge without necessitating a total revision of the theory.

This textbook was originally initiated under joint authorship with Pamela J. Brink, R.N., Ph.D., Associate Professor of Nursing, University of California, Los Angeles. As a contributing author, Dr. Brink has influenced the content and direction of the present text to a great extent. We collaborated in the development of the basic application of the model through joint lectures in a course entitled Basic Nursing Science at the UCLA School of Nursing, 1969–1971. Our thoughts were constantly challenged by the searching questions of our students. When we would fail to make clear the relevancies, they would be quick to demand clarification. The content also reflects the contributions of other faculty members who were involved in the development of the undergraduate curriculum at UCLA—Carolyn Carlson, Amelia Dowd, Ieva Kades, and Ruth Wu.

It is my hope that this text will stimulate others to provide alternative approaches to the care and understanding of human beings and their behavior. Just as it is important that people affiliate themselves with others to gain a sense of belonging and relatedness, it is also important that an individual also have a sense of relatedness to himself and to those ideals that provide meaning to his life. If this text can assist in the provision of individualized care and compassionate understanding of the wide variability of human behavior, it will have been successful.

I am indebted to the assistance of many people, some of whom have also influenced my life in very deep and personal ways. The constant understanding and support of Ms. Dorothy Johnson has been immensely important to me throughout this project. Mr. Albert Belskie, formerly of Prentice-Hall, Inc., and Mr. Harry McQuillen, editor, Prentice-Hall, Inc., have been sources of encouragement and unending support throughout the writing of the text. My warmest thanks are also extended to Ms. Eleanor Hiatt and the editorial-production staff of Prentice-Hall, Inc. who helped to refine the raw material into the book that is contained within these covers. Others who have provided important support include Bertha Unger, Superintendent of Nursing Service, UCLA Neuropsychiatric Institute; Frances Sharma; Hilde Kirsch; and Ann.

The unquestioning support of one's family is of utmost importance whenever an undertaking of this nature is begun. It is a formidable task to express the gratitude and appreciation and love that I feel for those of my family who have sustained me throughout the project. My parents have always encouraged my creative efforts, and my husband, Richard, and son, Joseph, have helped in every way possible to create the conditions that would permit me the freedom to create. Their belief in me has been boundless, and I hope that I have been able to repay in some small way their faith in my efforts.

Jeanine Roose Auger, R.N., Ph.D.
Nursing Service
UCLA Neuropsychiatric Institute

one | *Introduction*

Nursing shares with all health disciplines the goal of assisting an individual, or groups of individuals, to achieve a state of optimum health. This goal represents an ideal, a potential that may never be attained, partly because the criteria used to define and evaluate an ideal state of health are constantly changing. Evaluation of an individual's state of health is always limited by the present state of knowledge, which is expanded and modified by future scientific experiments. Furthermore, an individual is continually exposed to stress agents within the physical and social environment that require effective responses in order to prevent the development of an unhealthy state. This dynamic process of adaptation to stress precludes the development of an ideal state of health, even though a person may be in an apparent state of health.

Concern with the concept of health is relatively recent, compared to historical interest in conditions known as disease or illness. It is as if health were taken for granted and consequently ignored. In the past, the energies of man were directed toward understanding disease and illness, those catastrophic events that threatened the survival of man and society. Primitive belief was that disease was a state of possession by a spirit or devil as punishment for prior sins. A person was believed to have behaved in such a way that he or she had created the potential for illness. Consequently many religious and cultural ceremonies were designed to maintain a healthy relationship with the spiritual elements, and the ceremonies of healing were directed toward reestablishment of a connection with the spirit. This primitive view of illness still exists in many parts of the world and is exemplified in the use of sand paintings and other rituals by the American Indian in the treatment of illness.

Emergence of scientific, rational methods of observation and research resulted in identification of the microorganic and biological bases of disease. The condition of illness became a biological event, centered in the body. Interest focused on the physical variables of pathological processes and the etiological variables associated with various disease states. As knowledge has increased, it has become apparent that factors other than the physical can be significant contributors to the development of an illness. We now know that a patient must be regarded as a total human being, not merely as a physical organism affected by a variety of unknown pathogens. It is no longer sufficient to concentrate exclusively on the pathophysiological factors of disease. An orientation toward health also requires that the health professional be informed about the social factors that favor development of health problems, including overcrowding, poor sanitation, and poverty; the cultural factors which condition the life style of an individual, including food preferences and the symbolic meanings of life events such as illness, death, and birth; the psychological factors, including an understanding of perception and of the influence of past experiences on present behavior; and the environmental factors, including the effects of pesticides, oil spills, and air pollutants. In addition to the physical variables with which we are customarily concerned, any one of these major groups of variables may result in a disease process. Consideration of these multiple variables can result in more effective, efficient, and individualized care of patients, as well as in a higher level of health in the general population.

The traditional definition of health was a negative construct in which health was defined to be "the absence of disease." This conceptualization of health and disease as two mutually exclusive states was of limited value in describing the factors related to both. The demonstration of a state of health was possible only by verifying that a state of nondisease existed. Both the level of scientific knowledge and the degree of sophistication of diagnostic procedures available within a given setting would then become significant determinants of the criteria that would be applied to the measurement of a state of health. The circular construction of the definition—i.e., if health is absence of disease, then disease is absence of health—

does not bring one closer to a description of the components under-lying each state. The World Health Organization recognized the limited usefulness of this definition when it proposed that health be defined as a state of physical, mental, and social well-being, and not merely as the absence of disease or infirmity.[1] This conceptual-ization of health retains the implied, absolute character but broadens the range of variables which must be shown to be "healthy" be-fore an individual can be so labeled. The person who is physically healthy, and simultaneously impoverished due to economic depriva-tion would not be judged healthy within the limits of the World Health Organization definition.

Wu differentiated three classes of definition related to the con-cept of health, or wellness.[2] The latter term was proposed as a way of avoiding the connotation of health being directly linked to ill-ness. The three classes of definitions were:

1. wellness as the polar opposite of illness;
2. wellness on a graduated scale with illness;
3. wellness as a separate dimension from illness.

The two definitions discussed in the preceding paragraph are ex-amples of the first class of definitions, in which wellness is assumed to be the polar opposite of illness. The second class of definitions retains the assumption of opposites, but further assumes that the two poles are united by a continuum. A given individual may be placed at a specific point on the continuum that will represent the relative balance between healthy and nonhealthy factors present at the time of assessment. If the relative point of balance is toward the endpoint labeled health or wellness, then the person is judged to be basically well. An individual may move constantly along the con-tinuum, reflecting the dynamic processes underlying these two hy-pothetical states. An important distinction between the definitions of wellness as the polar opposite of illness, and those definitions that assume that the two states are linked by a continuum, is that the latter do not require the assumption that health and illness are mutually exclusive. Rather wellness and illness can exist simul-taneously within the same individual. It becomes possible to relate to both the healthy and unhealthy aspects of a person, rather than to deal exclusively with one or the other.

A primary difficulty with the health-illness continuum definition is that it requires the assumption of as many continua as there are dimensions of an individual in order to include all of the relevant factors that relate to health and illness. The consequence is to split the individual into many parts that are separate, rather than to consider him as a totality. For example, in order to describe the four primary dimensions characteristic of all individuals there would have to be a psychological health-illness continuum, a social health-illness continuum, a physiological health-illness continuum, and a cultural health-illness continuum. The interaction among these dimensions would not be considered in the evaluation of health. This problem is avoided by the third set of definitions that approach wellness as a multidimensional phenomenon that is independent from illness. These multiple dimensions interact with each other in a complex manner in order to create the unitary condition of wellness. This assumption of multidimensionality reconciles the splitting into parts that occurs when the definition is based upon continua.

For the purpose of this text wellness is assumed to represent *an integrated state of optimum function of the diverse biological, psychological, social, and cultural dimensions contained within the individual, which is manifested through behavior in response to the impact of complex environmental factors surrounding the individual.* Changes within one or more of the dimensions will result in an alteration of observed behavior. For example, a person may be physically healthy, but as a consequence of changes within the psychological dimension may require hospitalization for a behavioral disorder. Justification of hospitalization would be based upon changes in behavior that had been observed by both the individual and others in his environment. Such a person is clearly defined as unwell because of the aberrant behavioral changes. Less obvious is the situation of the person who manifests changes in behavior as a consequence of entering a new environment for which he has not developed a repertoire of adaptive behaviors. Attempts to use previously successful behaviors in the new setting may or may not be adaptive to the environment. Until the person is able to adaptively respond to the new setting he would be labeled "unwell," although not necessarily ill, within the framework of the proposed multidimensional definition. Although the altered environment may require changes in social behavior, it is possible through the interaction of the dimensions that the "unwellness" may involve behaviors that are psychophysiological in nature. Application of a multidimensional approach for the measurement of a unitary state of wellness implies that all aspects of the individual must be assessed before a judgment of wellness can be made.

An example will illustrate the basic differences between the conclusions reached by the health-illness continua approach and by the multidimensional approach. A 40-year-old woman, Mrs. G., was admitted to the hospital for possible acute hyperthyroidism. Measurements of the protein-bound iodine and blood cholesterol confirmed the diagnosis. Biopsy, however, failed to demonstrate any pathological basis for the increased secretion of thyroxin. Behavioral changes that had prompted Mrs. G. to seek medical attention included hoarseness, jitteriness, cardiac palpitations, and an annoying tendency to sweaty palms. She stated that she had felt well prior to the onset of the symptoms. Coincident with the onset of the symptoms was the hospitalization of her husband who had sustained a severe spinal injury. He had been given a favorable prognosis for complete recovery, although it would require an extended hospitalization. A few weeks later Mrs. G's son and wife moved in with her because the wife was severely depressed following the institutionalization of their retarded daughter. Although the two women felt that they were able to share responsibilities in the house without getting into conflict with one another, Mrs. G. still expressed the feeling that she had to be on her best behavior in order to avoid potential conflicts. The young couple were contributing toward the cost of maintaining the house; for which Mrs. G. was appreciative because her husband's injury and unemployment had created an extreme change in their financial situation, requiring them to accept public assistance. It was at this point that Mrs. G was forced to seek medical attention as the intensity of the symptoms continued to increase and she felt, at times, as though she were going to suffocate. She could give no reason why she might be feeling this way.

Analysis of the preceding example by the segmented approach of continua would require the use of four discrete measures of health. At the time of admission all four dimensions—i.e., physiological, psychological, social, and cultural—would be in the direction of illness. Following treatment and discharge from the hospital the physiological continuum would be restored to health, while the remaining three would have been unaffected by the direct treatment of the hyperthyroidism. Multidimensional assessment would demonstrate little change in the overall index of wellness for Mrs. G. unless the psychological, social, and cultural factors of unwellness were also modified. Thus, it can be seen that the condition of wellness is, for most people, a mixture of healthy and nonhealthy factors, the sum total of which describes an individual at a given time. It is for this reason that optimum health remains an ideal toward which some individuals come closer than others.

It is imperative that the health professional have an extensive knowledge of human behavior, including adaptive and maladaptive behavioral patterns. Often a nurse is able to identify complex maladaptive behaviors with a high degree of sophistication, whereas the ability to identify adaptive behaviors is less well developed. This variation in knowledge is a reflection of the traditional concern with pathological processes in nursing curricula. However, a nurse must understand adaptive patterns of behavior in order to help a patient develop healthy behaviors in those areas in which assistance is required, and to be able to discriminate when adaptive behaviors are present that should be supported and maintained. Unless the adaptive processes are known in depth it is not possible to know the maladaptive in depth, for they emerge from the same roots.

A systematic approach to the study of complex human behavior has long been the interest of philosophy and science. The biological and social sciences share the objective of describing, ordering, and explaining human behaviors with such accuracy that it becomes possible to predict the behavior of an individual, or groups of individuals, given a knowledge of the situation, the social rules governing such responses, and the personal characteristics and expectancies of the respondent. Although extensive research has been conducted in anthropology, sociology, psychology, anatomy, physiology, political science, and other behavioral sciences, the goal of predictability has yet to be attained.

A major source of the difficulty is the use of different explanatory concepts, or definitions, of behavior by each of the sciences. Even when the same terms are used, it is often with different meanings. The study of human behavior involves the description, measurement, and interpretation of those actions of a person which are *of interest to the observer*. This qualifying phrase—"of interest to the observer" —is of paramount significance. The purpose of the observation is a critical determinant of what behaviors will be observed, and of the types of people and settings in which the observations will take place. It is possible to conduct observations for discovery in which an attempt is made to observe everything that exists in the situation. However the observer must then do something with the information, or the material will be chaotic and meaningless. An anthropologist will observe the behavior of an individual, or groups of individuals, from a particular framework that serves to structure the observations and subsequent interpretations of the behavior in such a way that the results are meaningful to the science of anthropology. A psychologist, observing the identical situation simultaneously with the anthropologist, would have a different framework underlying his observations, and the resulting description of behavior would be different. Consequently, two descriptions would emerge from the same behavioral event, both of which might be valid, even though they would be dissimilar.

Even within the broad boundaries of the same field of study, the definition of behavior used by different investigators will result in disparate observations. The psychologist who ascribes to a behavior modification treatment of behavioral problems will examine behavior and its relationships to factors in the environment for possible contingencies. Behavior in this paradigm is restricted usually to overt, observable actions that can be manipulated through social and other forms of reinforcements. An analytical psychologist, representative of the other extreme, is primarily interested in the behavior of an individual as expressed through dreams, fantasies, and other material of a subjective nature that become available to the psychologist only through the willingness of the individual to share the information. The meaning of behavior for these two approaches has little else in common beyond the spelling of the word.

The term *behavior* can be applied to an extensive variety of phenomena, ranging from simple, isolated behaviors, such as the contraction of an isolated gastrocnemius muscle as a response to faradic stimulation, to more complex behaviors such as social aggression as a response to frustration. Unless the exact meaning of the word is stated, it becomes impossible to evaluate the results of research studies for possible areas of commonality and relationship to an overall unified framework of behavior. The process of defining behavior requires that the investigator describe in detail the nature of the phenomena to be studied including the proposed method of obtaining measurements of the behavior, the environmental conditions necessary for the presence of such behaviors, and the population of interest. This process of defining is called *formal operationalization*. A study of eating behavior, defined as the quality and quantity of daily food consumption by a tribe of cannibals, as compared to a group of post-gastrectomy patients, would identify areas of similarity and differences. Those factors that are similar in both populations are termed *commonalities* and represent shared characteristics that are assumed to be universal components of behavior. The areas of difference may represent *unique* factors that are specific to the particular population or setting studied, and as such may not comprise universal components of the behavior. The task of science is to operationalize these common and unique variables in such detail that a second investigator could duplicate, or *replicate*, the stated phenomena under identical conditions. This process of replication by an independent source serves to verify the reliability and validity of the original results. Once all the possible common and unique variables that contribute to the presence or absence of a particular behavior have been identified so that they meet the criteria of reliability, replicability, and validity it should be possible to predict *a priori* whether or not a behavior will be present under established conditions.

Establishment of the multiple variables that mediate behavior is perhaps the most difficult problem facing the behavioral scientist, for the human being is separate from, but dependent upon, his environmental surroundings and the conditions that elicit a particular behavior cannot be stringently controlled as in the case of most other subjects of interest, i.e., nonhuman species. It is therefore of great importance that the behavioral scientist consider not only the behavior itself, but as accurately as possible specify the conditions or situational factors present at the time the behavior was elicited.

The word *behavior* is used colloquially in three ways. The first usage is a description of an individual in terms of what he or she *is*. A patient may be described as sloppy, cooperative, hostile, or suicidal, or with an endless variety of other behavioral labels. The use of such descriptive labels is at best meaningless, and can be potentially quite harmful to the individual. What these labels are meant to state is that the patient is behaving *in a manner* that might be described as sloppy; this includes dirty, wrinkled clothing, failing to practice socially accepted grooming, and neglecting personal cleanliness. The patient who is labeled *suicidal* may be actively lethal and may be devising realistic plans for committing a successful suicide. Suicidal behaviors would include agitation, expressed feelings of sadness, refusal to interact with peers and professionals, and expressions of suicidal intentions. Whereas the label conveys ambiguous and stereotypic information, examples of the specific behaviors that resulted in the attachment of the label serve to communicate the current behavioral pattern of the individual. Such labels are useless in the scientific study of behavior, and should not be used by members of the health professions.

A second common use of the word behavior is to describe the actions of an individual in terms of the implied value associated with such behaviors. This includes those classes of behavior described as good, bad, normal, appropriate, and inappropriate. These value judgments often provide more information regarding the personal values of the observer and the conditions surrounding the occurrence of the behavior than they do about the person who is behaving. Such values tend to be expressed in absolute terms, as if some ideal, moral standard of behavior exists against which an individual can evaluate himself, or be evaluated by others. To value a behavior is to establish the observer in a position of judgmental authority over another. Why is one behavior good and another bad? Is the value of goodness the same for all individuals? for all situations? for the same person in different settings? Children are often taught that it is bad to behave in such a manner that another individual might be harmed or injured by one's actions. This valuation of "badness" is modified in the case of war, self-protection, or protection of others who are unable to protect themselves, including children and the elderly; in these situations an individual, instead of being punished for harming another, may be accorded the designation of hero. The behavioral actions of harming another are identical, whether it be in war or in intentional homicide; the difference between the two situations is that societal values differ by rewarding one situation and not the other. The application of value judgments in the study of behavior is of little value, although it does provide a useful standard for society at large.

The third use of the word behavior, and the one that will be used in this text, is more restrictive and includes *only those actions that are observable as responses to contingencies in the environment.* Verbal reports of an individual as to his affective states, perceptions, and thoughts may be included in this class of behaviors, but must be carefully evaluated for reliability and validity, and are completely contingent upon the cooperation and openness of the person.

It is imperative that some form of order be introduced to the analysis of behavior, so that those professions concerned with the behavior of individuals can understand and apply behavioral concepts. Nursing, for example, is intimately concerned with the care and support of individuals who are ill, or who are threatened with potential illness. The critical event of illness may disrupt the usual behavioral actions of an individual, and a nurse may need to assist that person to develop adaptive behaviors appropriate, or functional, in the situation. It is also important that those adaptive behaviors that have not been affected by the condition of illness be supported and maintained by the nurse. To be an effective therapeutic agent, the nurse must have a systematic and meaningful basis for assessing and evaluating the behavior she encounters. The conceptual model of Johnson provides an orderly approach to the understanding of complex behavior.[3]

The basic assumptions underlying the model are that:

1. the most reliable information source about people is their verbal and nonverbal behavior;
2. behavior, like any other discrete phenomenon, can be isolated for study, apart from the other human qualities, as an operational system.

Human behavior is viewed as those complex, overt actions or responses to a variety of stimuli in the environment that are purpose-

ful and functional to the individual. The apparent link between a stimulus and the behavior it elicits may not be obvious because of spatial or temporal factors. However, with the possible exception of autistic behaviors, which are responses to internally generated stimuli, it should be possible to identify the eliciting stimulus of the observed behavior. The practical difficulty is tremendous, for the ability of the human to memorize and internalize previously external stimuli creates the potential for extensive separation between the original stimulus event (Event A) and the associated response (Response A'). For this reason behavior may appear to be unrelated to the specific events occurring in the environment at the time observations take place.

The primary purpose of behavior is assumed to be directed toward maximizing a potential good for the individual and/or group. This potential good may be external, or it may be internal and consequently not immediately apparent to the observer. When an individual has received a message from the environment and has transformed the components of the message into a personally meaningful form, he must then determine whether an immediate response is appropriate, or whether the response should be delayed into the future. On the basis of this personal interpretation the individual selects from an extensive variety of possible responses the one that is predicted, or expected, to maximize the good. The choice may take place on a conscious or unconscious basis; this simply refers to whether or not the individual is aware of the decision process at the time the decision is being made. An important factor in the decision is prior experience with behaviors that have created a favorable result in similar situations, as well as experience with responses that have been partially successful or unsuccessful in the past. Since the selected response mode is based on the "transformed" stimulus event, which is individually determined, it is possible for two people to respond with diametrically opposed actions to the same condition, and each action will meet the criterion of maximizing the good. The factors that contribute to the development of personal interpretations of stimulus events are complex and will be discussed in Chapter 5.

Second, behaviors are assumed to be directed toward increased adaptation for the individual in relationship to the environment as well as to the specific situation. *Behavioral adaptation* refers to the dynamic adjustments to continually changing demands of both the internal and external environments designed to achieve a relatively stable condition of balance, i.e., equilibrium, between the two environments. *Stability* is characterized by relatively minor fluctuations centered around a hypothetical point of equilibrium, whereas *in-*

stability is represented by wide, irregular variations. Stable behavior, for example, would be characterized by constancy and predictability for a given individual, whereas instability would be demonstrated by ambivalence, hesitancy, and wide variability over time. The concept of *equilibrium* refers to a hypothetical fixed point of balance where opposing forces are cancelled out by one another. For example, body temperature reaches equilibrium when those factors that result in production of heat are precisely balanced by the factors that result in loss of heat. The point at which this theoretical equilibrium occurs may be anywhere from 97.6°F to 106.0°F The specific point of balance may vary between individuals as well as within the same individual. Measurement of body temperature levels for a large group of people will result in a normal distribution of temperatures, with the average temperature approximating 98.6°F Conditions such as extreme heat, fever, and brain damage may establish a new equilibrium point that is elevated above normal conditions. In terms of behavior, equilibrium refers to the theoretical point at which those behavioral components that inhibit behavior are balanced by those factors that facilitate the behavior. Evaluation of behavioral adaptation is determined in reference to a given individual and to the requirements of the specific situation. For example, eating breakfast at 7:00 A.M. may be a very stable behavior for a person, but would be highly maladaptive if the person were scheduled for surgery that morning.

Third, it is assumed that all human behavior is universal in nature, although the universality is obscured by cultural, social, environmental, and idiosyncratic influences, resulting in apparent uniqueness for each individual. Every infant at the time of birth and for a period of time in the postnatal period has the potential to develop learned behavioral patterns contained within any known culture, regardless of inherited characteristics. Tart has termed this phenomenon the *spectrum of human potentialities* and proposes that culture is the primary source of differentiation by selectively reinforcing certain potentials while simultaneously preventing the development of others.[4] Analysis of different cultural groups will demonstrate a group of shared, common, or cross-cultural behaviors, and a second group of unique, or culture-bound behaviors. Linguists have analyzed the vocalizations of 3-month-old infants and have reported that it is impossible to differentiate the babblings of infants on the basis of culture. Once the infant begins to make "intelligible" babblings, which are sounds that are recognized and reinforced by members of the particular cultural group as meaningful, the ability to make universal babblings is decreased, or extinguished. Eventually the patterned movements of the tongue that have been positively reinforced prevent vocalization not encountered within the native culture. Children's art themes have also been reported to transcend specific cultural groups and to demonstrate an organized pattern of development in which geometric forms such as the circle and straight line predate more complex figures. Ekman and Friesen have studied the universality of emotional expression and found that members of preliterate societies associate the same emotional concepts with facial expressions as do members of literate western cultures, providing further evidence to support the assumption that the nonverbal expression of emotion is a universal phenomenon.[5]

The universal behavioral components are initially transformed by the particular sociocultural factors, such as the shared social institutional structure, knowledge, beliefs, art, mores, and customs, that are shared among a specific group of individuals. These factors are learned by each member of the social group primarily through symbolic language and exert a powerful, but subtle, influence upon behavior. The common, universal behavior of eating is a relatively simple example. All peoples eat, but the differences in the behavioral components are widespread. A diet considered appropriate by one cultural group may be totally avoided by another group. The symbolic meaning attached to various foodstuffs is largely conditioned by cultural variables. Meat from a cow is rigidly proscribed for Hindus because of the symbolic significance of the cow in their religious belief system. Many religious holy days involve the expression of the symbolism of the ritual through special foods, be it the eating of eggs on Easter, fasting on Yom Kippur, or eating unleavened bread during Passover. Cultural variables may even determine who receives food when supplies are limited. In India, for example, when the availability of food is limited, male children will receive preference in foods over the female child, reflecting the cultural value which favors the male child.

Social roles and relationships further obscure the apparent universal basis of behavior. Status within a society includes certain expectations of behavior, as well as the avoidance of other specific behaviors. The group provides rewards for those individuals who are successful in performing the expected role behaviors, and provides sanctions and punishment for those who do not perform sufficiently well to meet the implied standard. Society, for example, provides certain guidelines for each individual as to the "accepted" way to eat or prepare food, and further states that everyone must feed himself (with the exception of infants and those who are ill). Eating food from others' plates is frowned upon and persons who engage in that behavior will tend to receive sanctions, e.g., a slap on the hand. The social rules tend to have greater influence upon eating behavior when the individual is in a group, as opposed to eating within his own personal setting or in isolation. Thus part of the means through which social rules modify behavior is through the physical presence of others.

The physical environment contains additional factors that significantly limit the universality of behavior. Recognition of these variables and their potential influence upon behavior is an outgrowth of the exploration of physical environments that differ from man's natural environment, such as the moon and the ocean depths. Behaviors that facilitate adaptation to the environment are transmitted primarily through the cultural modes. The effect of the environment on biological variables is also important. The negroid characteristics of high melanin production and flared nares are biologically adaptive to the hot equatorial environment and have permitted survival of the individuals within the particular limitations imposed by their physical location. In northern climates, fair skin coloring is an adaptation to the relative distance from the sun, and to the colder climate. A study supported by the Multiple Sclerosis Society is investigating whether the incidence of multiple sclerosis is greater in the northern half of North America than in the southern half. They found that the incidence rate of multiple sclerosis is 600:1,000,000 in Minnesota, whereas the rate for Mexico City is 16:1,000,000. Other questions include whether the presence of sanitation facilities significantly influence the incidence rate. Preliminary evidence indicated that as sanitation facilities have increased in Israel, the incidence of multiple sclerosis has correspondingly risen. One implication is that this illness may be a consequence of altering the environment.

Finally, the individual himself serves to obscure the universal basis of behavior. Evidence accumulated from studies of identical twins, who share a common genetic inheritance, support the importance of the environment in determining personality and other characteristics. An individual's past experiences, exposure to various situations, feelings, thoughts, expectations, and abilities are critical factors that serve to distinguish one person from all others and that provide the final blurring of these universal behaviors, transforming them into unique, personal events. It is this very individuality that prevents the easy prediction of behavior.

An individual, however, cannot divorce himself from the society and culture of which he is a member. In order to function successfully within society an individual must develop a variety of stereotyped forms of behavior that are approved and sanctioned by society. These stereotyped behaviors are typical, habitual responses that are consistent with the social convention governing the situation, i.e., the rules of etiquette. *Enculturation* is the process by which the knowledge of the group is transmitted to its members; *socialization* restricts the behaviors available to an individual while also providing for wide variations between individuals, so long as societal criteria are observed. These criteria range from the formal, legal structures which designate those behaviors that are harmful to society as a whole, together with the formal punishments associated with such transgressions, to the less formal designation of acceptable versus unacceptable behaviors. Society allows variations in behavior so long as they fall within the limits of normality, however broadly defined by the social group. Whenever behavior occurs on the border between socially defined normal and abnormal, members of the health professions may become involved in the modification of such behaviors. This involvement is contingent upon society's continued support of the profession and is controlled through formal, legal processes which define the area of expertise as well as the educational requirements for certification and licensing.

In order to remain viable, society must continually adapt to changing internal and external environmental requirements or pressures. The responsibility for the development of new expectations, and for the modification of already existing role behaviors, resides with the various professions that have been accepted as representatives of society as a whole. In 1974, for example, the American Psychiatric Association voted to remove the diagnosis of homosexuality from its list of psychiatric disorders. In effect they were stating that homosexuality, in and of itself, should no longer be considered by society to constitute a sexual deviation and that psychiatric treatment is appropriate only if an individual becomes anxious concerning his or her choice of sexual partner. This behavior, on the other hand, is still defined as an offense by the legal systems of most states. The situation, therefore, exists in which two professions have designated the same behavior to be simultaneously normal and abnormal. Eventually the discrepancy will have to be resolved, otherwise society as a whole will remain ambivalent and divergent in relationship to people who engage in such behavior.

Each individual is expected to develop behavioral patterns that are congruent with the basic expectations of the society in which he lives. A person may develop behavioral actions that are labeled as undesirable or harmful to society as a whole, or to specific individuals. Persons who emigrate to a new societal, or cultural, group must unlearn previously acceptable behaviors that are no longer appropriate to the new group, and also must learn new compatible behaviors in their place. The phenomenon of *culture shock* represents the period of adjustment to a new social pattern that differs from the previous one. To help individuals adjust to the demands of the group, society has accepted the health professions, together with educational and other social institutions, as the primary developers and modifiers of individual behavior. These professions assume the primary function of assisting an individual to develop and maintain behaviors in relation to the social expectations which apply, and if necessary, to help such persons modify existing behaviors in order to become more congruent with existing demands.

A crucial issue, however, is the importance of protecting the rights of the individual while simultaneously protecting society as a whole. There is always the danger that enforced conformity will result in a loss of freedom, such as George Orwell described in his novel *1984*.[6] A law intended to clearly demarcate the rights of the individual and the rights of society, in order to prevent potential abuse of freedom, was enacted in California in 1969. This law, the Lanterman-Petris-Short Act, states that a person cannot be admitted involuntarily for treatment in a mental facility because he "may do something dangerous" or is behaving in ways that are felt to be not in the best interests of himself or his family.[7] Only if the individual has threatened serious harm, attempted to harm, or actually physically injured himself or another person, and/or is declared to be unable to provide for himself, can that individual be detained against his volition, and this period of detention is limited to no more than 72 hours. Following that period a health professional must seek the approval of the courts in order to retain custody of an individual for treatment on an involuntary basis.

It is important that an objective behavioral measure be developed to describe an individual in terms of his, or her, unique place on the multidimensional system of wellness. This objective measure, in the format of an assessment interview, can be used by a nurse to evaluate the complex interrelationships among those multiple variables that contribute to behavior. Such a tool has the advantage of controlling the introduction of personal bias in the form of values and labels, but more important, it increases the accuracy of communication among members of the health team. Placement of a person on the continuum of wellness may also be done by the individual himself. However, not all persons are sensitive or aware of those behaviors that are potentially harmful, maladaptive, or inappropriate to himself or others. The health professional must then assist these individuals to become aware of the potential for change, the need for change, and the personal benefits that would result from the change process.

The focus of this text is on adult behavior, with primary emphasis on the general variables underlying behavioral actions. This includes a discussion of the biological, psychological, and sociocultural factors, and of their contribution to the observed actions of an individual in a particular situation. Analysis of specific behaviors includes an analysis of the situation and a knowledge of the general rules of behavior that govern such situations in order to determine whether or not an individual is behaving according to, or within the limits of, the rules. If the person is not meeting the expectations inherent within the situation then the observer must try to determine the source of the discrepancy between expected and observed actions.

The process of development from infancy to maturity is a major determinant of response capabilities. An infant has relatively few responsive actions available to meet a wide variety of situations. Those responses that are present tend to be global and nonspecific actions that serve a variety of purposes. As a child matures and develops a greater variety of responses, the responses become more specific to the eliciting event and more closely approximate the final, mature adult form. This text will deal with behavioral actions characteristic of the adult; the behavior of children can then be described in terms of the degree to which that behavior approximates the adult pattern.

This text will not emphasize psychopathology or other forms of deviant behavioral styles, although behavior modification, a therapeutic approach to the treatment of deviant or maladaptive behavior, utilizes a similar framework for the assessment of behavior. The interested reader is referred to the text by LeBow for the practical application of behavior modification techniques within the practice of nursing.[8]

SUMMARY. This chapter has introduced the student to the general problem of defining wellness and illness, and to the development of approaches to the study of behavior. The precise definition of the term is of vital importance to the evaluation of research and its subsequent application. Behavior is defined as those complex, overt actions or responses to a variety of stimuli in the environment that are purposeful and functional to the individual. The association between the eliciting stimulus event and the behavioral response may be obscured by spatial or temporal contingencies. It is assumed that the primary purpose of behavior is to maximize a potential good for the individual or group; that behaviors are directed toward increased adaptation; and that behavior is universal in nature, although the universality is obscured by cultural, social, biological, and personal factors. The systems analysis model of Johnson will be presented as an approach to the organization of knowledge regarding complex behavior.[9]

STUDY QUESTIONS

1. What is wellness? illness? How do these two concepts relate to one another?

2. How is the concept of behavior generally defined by anthropology? sociology? psychology? physiology? What are the similarities among the definitions? How do they differ?

3. Operationally define the behaviors exhibited by a patient who has been labeled as:
   a. cooperative,
   b. uncooperative,
   c. manipulative,
   d. dependent,
   e. hostile.

4. What provisions are made within a health setting to assist an individual to adapt to the expectations of the setting?

5. Observe a behavioral situation and examine the behavior in terms of those aspects that are shared by many individuals compared to those aspects that are unique to the individual.

two | *General Systems Theory*

Both ancient and modern man have attempted to understand and explain the world in which they live. Complex myths, philosophies, and religious beliefs were developed by ancient man as explanatory frameworks upon which an individual's relationship to natural phenomena could be established. Often these relationships were mediated through special rituals that were designed to create a feeling of participation with the phenomenon and reduce the apparent randomness and unpredictability of nature. The primitive myths, philosophies, and religions can be viewed as early precursory forms of modern theories which continue to influence modern man's relationship to the world surrounding him. The scientific age, however, has restricted their influence and has seen the development of theories, laws, and principles as explanatory concepts of natural phenomena. All scientific theories provide for the organization of knowledge regarding a specific class of phenomena in such a way that it becomes possible to:

1. examine the nature of relationships, i.e., interconnectedness, between the various multiple dimensions of interest;
2. make predictions regarding consequences of manipulation, or change, of one or more critical variables;

3. identify additional relevant dimensions of importance;
4. develop valid principles and laws that apply to the general phenomenon in specific conditions.

Theories, in general, are of practical use to both research scientists and professional practitioners. They provide a structure for observations of specific aspects of phenomena, and a guideline for interpreting those observations. In chapter 1, we referred to the influence of theory on research. A theory can dictate what phenomena will be studied and what phenomena will be excluded as irrelevant or inappropriate to the question of interest. Furthermore, a theory also provides the framework for attributing meaning to the results of an experiment; the meaning is determined by the manner in which the results relate to predictions based on the theory.

Professional practitioners also use theoretical frameworks in the process of observing, ordering, and explaining the phenomena with which they are involved. Engineers, for example, may apply theories from the fields of mathematics, physics, and geology in the design of a complex structure such as a bridge. Medicine and nursing apply theories from the biological and social sciences in the diagnosis and treatment of the unwell person. One of the greatest practical values of theories is their use in the process of decision making. A valid theory that has been extensively tested will provide a framework for predicting the potential effects of changing one or more dimensions upon the other dimensions of the phenomenon. For example, many states require a report concerning the potential environmental impact of major building projects that effect a change in the natural landscape. These environmental impact reports are predictions based upon current ecological theories of the consequences of environmental changes. Thus the application of scientific knowledge in the decision-making process can prevent, hopefully, subsequent irreversible changes in the environment.

Often decisions are based upon informal predictions, which are in turn based on an informal organization of knowledge about a person, a group of persons, or events. These predictions may be called *expectations*, based on the manner in which the person or object is predicted to behave in terms of the potential consequences of the decision. For example, children often develop many forms of behaviors designed to obtain an object that they desire based upon their expectation of how the parent will react to the request. If the expectation is that the parent will eventually give in and allow the child to have the toy or food, then the child will engage in behaviors that have been effective in the past until the object is attained. However, should the parent refuse to comply and instead punish the child's behavior, then the expectations have not been met. The major difficulty with decisions based upon these informal theories, or expectations, is that the basis of the decision is untested. Instead the decision is based on assumptions regarding the manner in which another person or situation will operate under the established conditions. Any change that occurs may or may not be a consequence of the decision that was made, and as long as the decision is based on untested, informal knowledge it becomes impossible to evaluate the nature of the change. Nursing has operated primarily on such an intuitive level in the development of specific interventions for patient problems. A major goal of the scientific movement within nursing has been to develop a formal, explicit theoretical framework that can be used for the development of formal decisions and predictions. This requires extensive testing of the many assumptions about patients and their behavior that have served as the basis for decisions on care and treatment.

General systems theory represents the organization of knowledge concerning the world or the specific phenomena of interest as a complex system where the focus is on interaction among the various parts of the system rather than on describing the function of the parts themselves.[1,2] This theoretical framework is a general structure out of which has evolved an extensive number of theoretical approaches to the understanding of physical as well as social phenomena. Educational systems, game theory, cybernetics, system engineering, information theory, computer sciences, decision theory, and communication theories are specific examples of the application of the principles of system analysis. This approach has also been applied in psychiatry, particularly in family therapy, as a basis for therapy. Theories based upon the systems approach are not in conflict with other forms of theoretical frameworks, since the knowledge concerning the function of the isolated parts has been derived from these earlier approaches. But the primary emphasis on interaction results in a dynamically oriented theory, as opposed to a mechanistic descriptive approach. A major advantage of this theoretical approach is that decisions concerning the potential consequence of change within one portion of the system can be more readily related to distant parts of the system because of the focus on interaction, boundaries, and energy exchange. This chapter presents the basic concepts of general systems theory which form the foundation for discussion of the behavioral systems theory of Johnson, to be presented in the next chapter.

Systems analysis simplifies and orders discrete empirical phenomena into abstract units that are amenable to analysis. Once ordered, these units become theoretical constructs that can be examined *as if* they were totally isolated from their natural environment. A *system* is defined as a whole with interrelated parts, in which the parts have a function and the system as a totality has a function. This view is expressed in gestalt psychology by the principle that the function of the whole is greater than the sum of the function of the parts. Systems operate in relation to both immediate and distant environments through the processes of input, transformation, and feedback. Disorganization is always present in a system since the parts of a system must constantly adjust to the continuous input, and also because there is always some energy loss related to the process of transformation and subsequent feedback. Systems, however, tend toward balance, or equilibrium, among the parts and processes in order to promote survival of the system in totality.

It is possible to characterize various systems on the basis of their degree of interaction with the surrounding environment. Systems may be classified as *open* or *closed*, depending upon the degree of porosity of the boundary-maintaining mechanisms, with the more open systems resulting in greater contact with the surrounding environment. All living systems are *open systems*; the term *closed system* is used in a relative sense, since the demonstration of a totally closed system is unknown at this time. The *environment* surrounding a system is defined as those elements external to the boundary of the system which correspond to the space-time continuum of interest. The more immediate the environment—i.e., the closer the proximity to the system under analysis—the greater the influence on the system and the greater the proportion of total input to the system. Conversely, those elements more distant will exert lesser influence. When the above elements and processes have been identified for each system, then a formal organization of ideas or relationships can be developed, and the rules that appear to apply to the organization of the system can be formulated.

Theoretically all natural phenomena can be studied under a systems approach. Systems can be organized in terms of hierarchical levels of complexity and the simpler systems may constitute subsystems of a more complex system. For example, the structural components of a cell can be studied as independent systems with specific functions, or as subsystems of the entire cell. At a higher level the cell comprises the subsystem of an organ, and so on. A person is a subsystem of the family, which in turn is a subsystem of the community, which is a subsystem of the state, which is a subsystem of the country, and so on. This level of complexity can be extended to the universe as a system, and even then the universe may in fact be a subsystem of a more complex system. It is important to be consistent as to the level of the hierarchy with which the system is concerned, since all components must be of the same complexity. The following discussion presents the basic concepts of the systems approach.

DEFINITION, STRUCTURE, AND FUNCTION

A *system* is defined as a whole with interrelated parts. Any portion of observed reality can be defined as a system so long as it is possible to discretely isolate it from its surrounding environment for analysis. A stone represents a relatively simple system on the dimension of complexity, as compared to the engine of an automobile, or the behavior of an individual. Nevertheless, all three phenomena constitute systems within the criteria of the definition given above and can be analyzed according to the systems analysis approach.

Once the system of interest has been clearly defined in an abstract maner, it becomes possible to specify the structure and function of the system. *Structure* refers to the arrangement and nature of organization among the parts of the system, while *function* refers to the dynamic interaction among the component parts. The parts, in interaction, form a gestalt which reflects the total system function at a given point in time. In a smoothly functioning system it should not be possible to discriminate the function of one part from another.

An example may help to clarify the relationships among these three concepts. The gastrointestinal system can be defined as the physiological system concerned with ingestion of fluids and foodstuffs, their subsequent digestion, and elimination. This system can be isolated from all other physiological systems of the body, such as the nervous system or skeletal system, for the purposes of theoretical analysis, although within the total person all of these systems are interrelated and affect the function of one another. This process of isolation, however, is an attempt to discover the rules that govern the function of the gastrointestinal system. The analysis of function must account for the influence of these other systems as they modify the activity of the system under examination as components of the surrounding environment. The structure of the gastrointestinal system includes the mouth, tongue, teeth, esophagus, stomach, gall bladder, pancreas, small intestine, large intestine and the various sphincters. A major function of the system is the regulation of food ingestion. This function involves activity of the total system, although some parts contribute more directly to the ingestion of foods, while others are primarily involved in the function of digestion and/or elimination. In order to determine the contribution of each part of the system to the total function it is necessary to develop various strategies for isolating the various parts from each other. For example, in order to differentiate the role of oral factors from that of gastric factors in the amount of food ingested at a given time it is necessary to artificially disrupt the natural connection between these two parts of the system. Creation of an esophageal fistula permits the study of the influence of food bulk within the mouth as separate from the factor of food bulk entering the stomach. Application of the systems analysis approach makes it possible to discriminate the contribution of the various parts of the total system in a purposeful way.

The function of a particular system may be viewed either in its relationship with other systems of an equivalent level of analysis, or in relation to the greater macrosystem of which it is a subsystem. The gastrointestinal system, circulatory system, and musculoskeletal system represent equivalent levels of analysis. Together these isolated systems constitute the parts of the macrosystem of physiology. When analysis is performed between systems of equivalency it is possible to relate the total function of one system to the other. But analysis of the relationship between the physiological system and the circulatory system requires that the latter be treated as a subsystem since it constitutes a part of the whole. Each of the subsystems of the physiological system, then, may be examined as a whole with interrelated parts that have a function specific to the subsystem; a function in relationship to the other subsystems; and a function in relationship to the physiological system as a whole.

BOUNDARIES

Once a system has been ordered into its discrete, empirical phenomena, and the system has been isolated from its particular environment, then the boundaries between the system and its environment can be established. A boundary may be defined as a more or less open line forming a circle around the system, where there is greater interchange of energy within the circle than on the outside. It is helpful to visualize a boundary as a "filter" which permits the constant exchange of elements, information, or energy between the system and its environment. Any system which is living, and consequently requires input of nutrients to prevent energy store depletion, must by definition be an open system with environmental exchange. The more porous the filter, the greater the degree of interaction possible between the system and its environment. When the boundary is relatively impermeable to input, the system becomes relatively isolated from its environment. Such boundaries may be termed dynamic boundaries since the porosity of the filter can vary over time in relationship to the status of the system, thereby allowing expansion and contraction of the mechanisms which govern energy interchange. The ability of a boundary to control the degree of interchange is of great importance in regulating the amount of the environment that is allowed to impinge upon the system at any time. If input from the environment is excessive and exceeds the limits of normal system processing, the boundary can contract, thereby reducing the level of environment impingement.

Boundary function also relates to the predictability of a system. The more open a system and the more porous the filtering system, the greater the amount of input received from the environment. When a system receives many inputs it becomes less possible to specify the exact source of the system response to the environment; consequently the system becomes less predictable. Conversely, the more closed a system is, the less input will be received from the environment, resulting in a more predictable response of the system. A rock represents a closed system in which the boundary is relatively unaffected by the surrounding environment. It is possible to predict with close to one hundred percent accuracy the response of a rock to erosion by natural forces, versus the response of a rock to a rock crusher. At the opposite extreme the human being represents an open system that can receive and respond to a multiplicity of environmental stimuli at a given moment, and its response is less predictable to an observer.

INPUT, FEEDBACK, TRANSFORMATION

Since all systems have boundaries that distinguish the system from its environment, and since all boundaries are considered to be "filters" that limit and direct the amount and type of interaction the system will have with its environment, systems analysis must also include the processes of input, feedback, and transformation. All open systems must receive input, in the form of communication or physical energy from the environment, which allows the system to function in phase with its environment and to detect any disparity between current level of system activity and the ideal. The system may distinguish input that is useful for the survival of the system from input that is nonuseful or harmful. Some forms of input may be immediately used by the system, whereas other forms must undergo a process of transformation into functional or meaningful forms. The administration of simple sugars through an intravenous preparation directly into the circulatory system is an example of environmental input which can be immediately utilized by the physiological system as a source of energy. A meal of carbohydrates must undergo a complex, chemical degradation process, called digestion, by which it is transformed into biologically useful foods. Through the sensory modalities of vision and hearing, an individual receives information about the environment that may be transformed into knowledge. This input or communication may be stored in the form of memory for use at a later time, or it may be utilized for an immediate response.

A system not only takes in information and energy from the surrounding environment, but it also provides output to the environment in similar forms. The energy output creates the process of feedback, which helps to promote system stability. The system must be able to detect the presence of any disparity between actual system function and "ideal" system function at a given time. Corrective action cannot occur unless a given source of disturbance is successfully discriminated from other possible sources, otherwise the system must use nonspecific forms of responses to deal with the disturbance.[3] Negative feedback is the process by which the system continually adjusts its level of activity in terms of the measured difference between the ideal and actual. It is termed *negative*, because the process is intended to decrease the difference between these two points.

DISORGANIZATION AND EQUILIBRIUM

Because systems are continuously involved in the processes of input, transformation, and feedback, they tend to be in a constant state of disorganization or disequilibrium. The parts of a system are not perfectly integrated. Consequently there results a certain degree of tension or stress within the system. Systems, however, strive toward equilibrium or balance in order to reduce the tension as much as possible. For a system to survive it must exist in a relatively steady state in which the input, feedback, and transformational processes are in relative harmony, with no one process assuming a greater portion of the energy available to the system. Excessive input is not useful to a system if it cannot be transformed for use; rather it places demands upon the system for reduction of the excessive levels and decreases the efficiency of the overall function. Hyperventilation is an example of the decreased efficiency of the physiological system as a consequence of excessive input. Hemorrhage is a form of excessive output from the circulatory system to the environment, which places extreme demands on the energy of the system in order to replace the loss. Insulin shock represents insufficient input combined with excessive transformation, whereas diabetic coma may be viewed as the consequence of inadequate or insufficient transformation of energy products.

A system that is in relative equilibrium responds to external stress by resisting the influence of the disturbance, unless it is of sufficient intensity to disrupt system balance. Resistance to a disturbance may take the form of not acknowledging its existence; of activating homeostatic forces that immediately restore or recreate a balanced state; or of reaching a new point of equilibrium. It is necessary to identify those forces that promote change and those that resist change in order to describe the process of resistance to stress.

ENERGY

All dynamic, open systems require continuous supplies of energy in sufficient quantity so that all demands for system integrity can be met. This constant source of input of energy from the environment takes place in forms which can be stored and converted into energy as required by the system, as well as in immediately available forms. At the same time, energy output from a system serves to maintain a sufficient level of energy in the environment which can eventually be reabsorbed by the system. Essentially, the system and its environment form a symbiotic totality, in which each serves as a source of energy for the other.

Each system requires a specific form of energy for its continued function. It is necessary for the system to have some provision for the conversion of available potential energy into appropriate forms that can be utilized on short notice. This conversion process itself requires energy. For an open system to survive, energy must be maintained in adequate amounts to prevent total disintegration. The provision of energy must be sufficient to account for that theoretical energy or heat which is contained within the system, but which cannot be extracted for work. This unavailable energy, *entropy*, is in part the basis of all disorganization present within a system.

The most important factor governing the rate of energy transformation is the level of ongoing activity within the system. If the activity level is high, utilization of large quantities of energy must be compensated for by an increased rate of energy transformation in order to meet current demands. If the system is relatively quiescent, the rate of transformation can be comparably reduced. Certain types of activities, such as physical exertion, exercise, and strong emotions, demand an increased rate of energy transformation to sustain the activity for a sufficient time to ensure survival of the system. During such activities it is possible to develop a temporary deficit of energy stores with little difficulty, since unless survival is ensured the system as a whole will become extinct. The tolerance of deficits varies greatly. The central nervous system, for example, is extremely sensitive to deprivation of oxygen and glucose, and any disturbance of energy input to the brain can be tolerated for no longer than three minutes without critical problems, and possible death. At the other extreme, people may be deprived of contact with other humans for days with few changes in behavior, but even then there is a finite limit to the ability to tolerate social deprivation with no effects.

Each system must have a provision for the storage of energy in a form which can be immediately utilized during excessive demand periods. The ability to store energy provides the system with a degree of independence from the environment since these stored forms can be utilized before the system is required to seek energy from the environment. Sugar and fats stored in inert forms can be transformed into glycogen when required, thus enabling the organism to engage in an activity without constantly being supplied with sugar.

At a given moment there are five different forms of energy in a system:

1. energy that is entering the system in a nonfunctional form;
2. energy that is in the process of being transformed into a functional form;
3. energy that is being utilized;

4. energy by-products that are being dissipated to the environment;

5. entropic energy.

The interaction between the first four forms of energy takes place
at a relatively constant rate, resulting in an apparent steady state.
Disruption of system equilibrium occurs when there is an imbalance
in the relative distribution among the four energy forms. Food consumed during a meal represents a relatively inert form of energy that
must be broken down into its basic chemical constituents before it
can become available to the system. This initial process, whereby
food is transformed from a nonfunctional form into a functional
form, is termed *digestion*. The resultant chemical components are
then used by the individual cells as a source of energy to maintain
functional integrity and activity of a particular cell. This second
process is termed *metabolism*. A nonfunctional end product of the
process of metabolism is heat, which must be dissipated through the
skin to the environment in order to maintain system integrity. It is
apparent that the creation of an imbalance in any one of the three
active processes will result in a problem that could disrupt the integrity of the system. Any event that diminishes the quantity or quality
of food ingested by an individual, such as starvation or anorexia
nervosa, restricts the supply of energy readily available for utilization and results in the depletion of inactive forms stored for use during increased periods of energy consumption. Such individuals will
fatigue rapidly when participating in any activity which increases
the rate of metabolism, because they have inadequate energy supplies. Any factor that increases or decreases the metabolic rate will
be reflected in a compensatory change among the other two processes. For example, hyperthyroidism represents an increased rate of
metabolism that subsequently involves the generation of increased
heat within the body, resulting in a low grade fever. Secondary to
the increased metabolism is an increased requirement for the ingestion of foods.

The specific form of energy is determined by the structure of
the system and its component parts. The biochemical forms of energy
that sustain physiological function have been extensively identified
and studied. The energy that sustains the psychic processes, such as
cognition, perception, and problem-solving, has been termed *libido*,
but the form of this energy has not been demonstrated empirically.
The energy that sustains a group of individuals is generated through
interaction, but is greater than the sum of energy of the individual
parts or members. A highly energized group, such as a mob during

a riot, can literally generate an explosive tumult which carries the individual along effortlessly. It is important to remember that the various parts of the system, or subsystems, may require very different forms of energy, and it is necessary to account for the different forms of energy and their interaction.

Energy distribution within the system is determined by the level of activity of the various components which comprise the system— i.e., the law of supply and demand. Each component receives a baseline level of energy that enables continued function; then those components that are most active at a given point in time receive the greater amount of the remaining available energy. Equilibration does not mean that the energy is distributed equally among all active and inactive components; rather it refers to the equitable distribution of energy in view of the current level of activity within the various parts. When a person runs, the blood supply must be shifted from the internal viscera to the skeletal muscles, rather than being evenly distributed throughout the entire circulatory system. Those components receiving the greatest amount of energy will be the most dominant determinants of overall system activity at that time. Each component, however, contributes to the overall outcome, although the contribution of the less active ones will be less apparent. Excessive or continued dominance by one part over all others may eventually place the entire system in jeopardy, if one component threatens to usurp the total energy supply to the system.

SUMMARY. General systems theory includes the examination of a system as composed of parts with specific structure and function. These parts are bounded by an imaginary line that serves to separate the system from its surrounding environment. Information in the form of physical energy or communication may transcend the boundary in order to enter the system. The system must then transform the energy input into useful forms that have meaning to the system under consideration. In turn, the system produces energy which is fed back into the environment. Negative feedback allows for the constant adjustment of internal system function. Systems tend to favor a state of equilibrium or stability which represents a balanced state of energy distribution in terms of the level of activity of the component parts. Disruption of stability results in stress to a system, which necessitates an adjustment in function and energy distribution.

Chapter 3 will present behavior in a systems analysis framework. It is recommended that the list of suggested references be reviewed for pertinent sources of additional detail concerning the basic systems approach.

STUDY QUESTIONS

1. Select a simple system, such as a cell, and describe the system in terms of:
    a. the parts, their structure and functions.
    b. the nature of the boundary.
    c. energy sources.
    d. conditions that disrupt stability of the system.
2. What are the consequences of stress on system function? boundary porosity? energy distribution?

three | *Behavior as a System*

Nursing practice has focused, historically, upon the care of the individual who is ill or threatened with potential illness, and has emphasized consideration of the particular needs of the individual. In contrast, nursing science has followed the traditional assumptions of the nomothetic approach to research that underlies most of the social and physical sciences. The idiographic and nomothetic approaches to behavior represent very different positions concerning the study of individuals and their relationship to others. The *nomothetic*, or group, approach assumes that a particular characteristic, or set of characteristics, is universally applicable to all individuals. Differences between specific individuals are assumed to represent their different locations on the dimension of interest. The definition of health, for example, which places two persons at different points on the continuum of health and illness reflects a group approach; it is assumed that the characteristics of the dimension apply to all persons and that any individual differences are indicative of variations among the relevant characteristics. The *idiographic* approach, in contrast, emphasizes that individuals differ not only in the way in which the characteristics vary, but further that they may have totally different dimensions or relevant characteristics. The multidimensional definition of wellness does not, for example, define the specific characteristics that comprise the state of wellness but assumes that a wide variety of characteristics relevant to the individual combine in some fashion to create a sense of wellness.

A major difficulty with the idiographic approach to behavior is that the approach does not formally specify common characteristics, and there must be as many theories as there are individuals in order to account for the apparent uniqueness of each person. Examination of human behavior, however, discloses that both approaches are appropriate, depending upon the particular interest of the practitioner.[1] There are many areas of commonalities among individuals that can be the basis of a nomothetic theory of human behavior with wide applicability to groups of individuals. Most, if not all, people eat, drink, sleep, walk, run, laugh, cry, and interact with others. The performance of these behaviors may be highly unique to each individual under certain circumstances such as in personal settings, whereas the same behaviors may demonstrate a high degree of shared performance when exhibited in more collective settings. Nursing must consider both the shared and the unique characteristics of the individual, i.e., an integrated idiographic-nomothetic approach to both practice and research.

The Johnson model represents such a combined approach.[2] The model specifies universal structural components to behavior; the specific behaviors, however, are unique to the individual and cannot be specified *a priori*. It is essential that the conditions that are directly related to the occurrence of a behavior be isolated from those factors that may occur simultaneously, but which are not related to the observed event. The process of isolating behavior as a quality separate from all other qualities of the individual is intended to develop an abstract formulation of behavior in order to identify the areas of commonality and uniqueness. Behavior can be viewed as a complex system with a surrounding environment that operates according to certain rules that can be specified.

THE BEHAVIORAL SYSTEM

The Johnson model of behavioral system analysis assumes that behavior is a quality of the individual that can be separated from all other qualities with which it is normally found. The model examines the behavior, not the individual who may be behaving at a given time. Therefore the behavior is the system, not the individual. The system may be defined as *those complex, overt actions or responses to a variety of simuli present in the surrounding environment that are purposeful and functional.* These actions or response modes can be grouped into a variety of subsystems; the model assumes that there are eight basic groupings of behavior that can be distinguished in terms of the purpose or function of the behavior. One assumption of the model is that man is a bio-psycho-social being situated in a particular cultural setting. These biological, psychological, and social qualities comprise part of the surrounding environment of the behavioral system. A second assumption of the model is that behavior is an integrated response to an external or internal stimulus, which is modified by the environmental factors of the behavioral system, i.e., the physiological, psychological and sociocultural regulatory factors. Behavioral responses, therefore, are dependent upon the activity level of each of these regulatory groups, but as in any integrated system the level of activity of the behavioral system is higher than the level of activity in the surrounding environment.

Behavior functions primarily as a response modality and is the primary mechanism of communication between an individual and the surrounding environment. As such, the behavioral system involves all of the observable actions of an individual over time, which constitutes the output to the environment. Behavior is also conceptualized as the integrator of the response of an individual to constant input from the environment; it is purposive, or goal directed, in relation to the requirements, drives, and/or perceptions of the individual.

STRUCTURE OF THE BEHAVIORAL SYSTEM

Johnson defines a behavioral system as constituting a complex of observable features or actions that determines and limits the interaction of an individual and his environment and that establishes the relation of the person to the objective events and situations in his environment.[3] She has specified eight parts of the behavioral system, i.e., subsystems, each of which is assumed to be a universal component. The eight subsystems are titled:

1. Ingestive
2. Eliminative
3. Dependency
4. Sexual

5. Affiliative
6. Aggressive-Protective
7. Achievement
8. Restorative.

Each subsystem is composed of structural components that interact in a specific pattern. The parts are:

1. goal
2. set
3. choice
4. action
5. sustenal imperatives.

Goal

The eight subsystems of behavior share the general goal of the whole system, that of ensuring survival and adaptiveness to the environment. The *goal* of a behavior is the purpose for which the behavior is intended. It was stated that the overall goal of behavior limits and directs the interaction between the individual and the environment in such a way that the possibilities of survival are maximized and the individual can obtain the desired outcome. The goal of a specific behavior may not be apparent to either the observer or the behaving individual. *Manifest goals* are the immediate intentions of the behavior, as compared to *latent goals* that must be inferred on the basis of observational material and prior behavior.

The dimensions of time and space influence the properties of the goal. Temporally, some behaviors are determined by the immediate goals, which are obtained after a relatively short time interval following completion of the behavioral action. Other behaviors are determined by more distant, long-term goals. Differentiation between the two temporal contingencies is not always immediately apparent. A child may practice his violin after school to please his mother and avoid possible punishment (short-term goal), or to become a concert violinist (long-term goal). Both goals may be functioning simultaneously, in which case it is necessary to identify the primary goal in the immediate situation and consider the other goal as a secondary influence. The particular situation or space in which the behavior occurs further affects the specific goal. If the desired outcome is immediately available in the setting the behavior will occur with greater consistency than if the desired outcome is not contained within the setting. A person is more likely to come to work on payday, to receive his wages, than on the day following.

Each subsystem has specific goals that are defined in terms of the primary function of the subsystem.

1. Ingestive goal: To bring into the individual a substance, object, or information that the individual perceives or determines to be lacking. This goal of "taking-in" may be for pleasure, gratification, relief of pain, knowledge, or safety.

2. Eliminative goal: To release, let go, get rid of waste products, excess or nonfunctional matter within the system. It may be viewed as a goal of tension-reduction.

3. Dependency goal: To seek help to obtain another goal, or to seek assistance in a task-related activity.

4. Sexual goal: To procreate and ensure survival of collective individuals, or to obtain pleasure from sexual activities. This goal includes the development and maintenance of an adaptive sexual identity for the purposes of seeking or attracting a love object.

5. Affiliative goal: To belong to or be associated with others in some form of specific relationship. This goal includes the process of interaction.

6. Aggressive-protective goal: To protect oneself, others, or property from real or imagined harm, or threat of harm in the form of attack.

7. Achievement goal: To master or control oneself and the environment in such a way as to obtain a desired object, position, or need.

8. Restorative goal: To maintain energy balance throughout the system through transformation and redistribution of the available energy throughout the system, in accordance with the demands of the various subsystems.

Each of the above goals is conceived of as the *boundary* for each of the eight behavioral subsystems. The subsystems interact with each other as well as with the surrounding environment. At a given time they can be ordered in terms of a hierarchy, from most active to least active. The most active subsystem will define the observed behavior, even though all of the subsystems are active to some degree. Figure 1 illustrates the interdependence of these eight subsystems and the concept of unique and supplemental functions. Each subsystem has a number of response tendencies or actions that are clear-cut and specific to the goal of that subsystem. The behavior of eating is an immediate response to the perception of hunger, for example. The subsystems also share a pool of supplemental responses which can be used by any number of subsystems in order to obtain the desired outcome. An individual may appear to engage in actions associated with the dependency subsystem, such as requesting help, as a means of meeting the goal of affiliation. These shared responses, or supplemental pool of response tendencies, permit more economical use of behavioral actions. On Figure 1 the white tip of each of the eight petals represents the unique response pool associated with the subsystem, and the overlapping areas represent the supplemental pool of response tendencies.

Set

The second structural unit of the behavioral system is that of *set*, or the consistency of response to certain environmental stimuli. Set contains two components: *perseveration* and *preparation*. Perseveration, or the *perseveratory set*, refers to habitual responses, in which a behavior has become so consistent that the individual is no longer aware of the action unless it is called to his awareness either by another individual or by alterations in the situation. The perseveratory set represents largely routinized patterns of behavior that are repetitive. The point at which a person brushes his teeth during his routine morning personal hygiene may be fairly consistent, even though the individual is unaware of the patterning. The person driving back and forth to work every day over the same route may find that he has driven miles before he becomes aware that he is driving. In another case, an individual may find himself driving to work on his day off, when he was actually intending to go in the opposite direction. All these examples demonstrate behavioral responses of a habitual, perseveratory nature in that they occur in response to situational conditions in the environment, but without total awareness on the part of the respondent. The perseveratory set is modified by individual temperament, past experiences, the level of general distractability, and learned social customs such as etiquette and beliefs.

The function of the perseveratory set is enhanced by the prin-

Figure 1
Behavioral Systems Balance

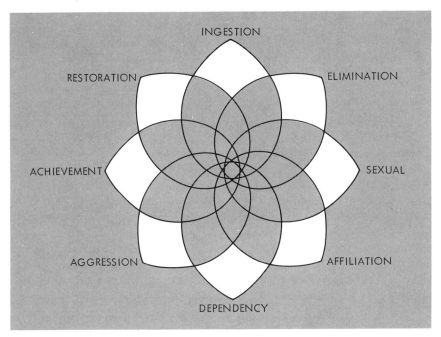

ÇIples of economy and inertia. The *principle of economy* involves the rate of energy expenditure, or conservation of energy. Essentially, it requires less effort, less energy expenditure, to repeat a learned behavior than it does to develop a new behavior through trial and error. The behavior involved in eating—putting food on a fork and placing the fork in the mouth—becomes easier and requires less energy consumption if it is continued in the way it was first learned, rather than if a person always has to attend to the specific behavioral actions. Americans often become aware of their particular pattern of holding eating utensils only when they travel through countries where a different pattern of behavior is used for conveying food from the dish to the mouth.

Related to the principle of economy is the assumption that behavior will be stable, given order and predictability in the person, the situation, and the environment, and that such behavioral stability utilizes less energy. The *principle of inertia* reinforces the principle of economy, and means that once a pattern of behavior has been learned it will remain stable and fixed as long as the pattern remains adaptive. An individual is reluctant to learn another method and to give up the easiest method of responding to a situation unless some external motivator is applied, or unless it is demonstrated that the existent response pattern requires more energy than another pattern to accomplish the same goal. For example, once an individual has learned to write with his right hand, he is unlikely to spontaneously begin to write with his left hand, because it requires too much effort and expenditure of energy to do so. Further, if a behavior is initially learned incorrectly it will require greater energy to correct it than to learn a new behavior as if no behavior had yet been established. A person must not only learn the correct components of the behavior, but must concentrate on excluding the incorrect ones. Someone who has learned to play the piano with incorrect fingering will experience much greater difficulty in developing correct fingering technique than will a novice who learns the technique correctly the first time. The increased energy expenditure needed is a consequence of the prepotency of the perseveratory set.

The *preparatory set* is contingent upon the function of the perseveratory set. The latter set will be dominant in familiar, everyday situations for which adaptive responses are already established. The preparatory set, in these conditions, supplements the function of the perseveratory set through a process of selective inattention which enables the individual to attend to the critical stimuli and exclude the irrelevant stimulus inputs. At any given moment an individual is deluged with a wide variety of sensory stimuli that originate from both within and without the system. The preparatory set functions to establish priorities for attending or not attending to the various stimuli. There are times when the mother of an infant does not "hear" the cry of her child, while at other times a cry will evoke an immediate response. The mother's close relationship and past experience with the child has provided her with a set of selective inattention to certain types of cries. In this way the mother can do other things and respond to other stimuli existing in the environment. A visitor, on the other hand, who is new to the situation does not have the same selective inattention and will consequently hear every cry the infant makes.

The mood of an individual on a particular day may also affect sensitivity to various situational stimuli. The familiar story of the man who has had a fight with his wife and who then takes offense to what others say to him, is an excellent example of the way in which the mood of an individual will affect his preparatory set. Whether an individual has just received a compliment or a criticism, whether he has just engaged in a disagreeable activity such as paying income tax or a traffic fine, are all factors which predispose an individual to view a subsequent situation in a particular way.

The more unique a situation is, the less an individual is able to rely on his previously established perseveratory set. Behaviors learned within a specific situation can be generalized to a wide variety of situations, as long as there are factors shared with the original setting. If an individual encounters a totally unique situation, for which he has developed no previous response pattern, he is totally dependent on the information obtained via the preparatory set for determining the possible class of response that would be adaptive in the new situation. An individual who is hospitalized for the first time and who has never experienced illness before has been placed in a situation in which his perseveratory set provides little or no assistance; rather he must utilize extensive amounts of energy to respond to each new situation he encounters. Predictability has been eliminated from the environment, and the individual's behavior will appear disorganized, unstable, and ambivalent until an adaptive set has been established. Another individual who has developed a set through prior hospitalizations will require less energy in responding to a variety of situations encountered within the hospital setting. Moos suggests that indiscriminate use of consistent behaviors over a wide range of situations that are dissimilar in content tends to be characteristic of the severely disturbed, less mature person.[4] Healthier individuals will demonstrate less consistent behavior as a result of being more sensitive to the subtle differences between situations.

Choice

The third part of behavior involves *choice*. Choice refers to the process of selecting from available alternatives the specific behavioral response that will best meet the goal and attain the desired outcome. An individual acquires the ability to generate vast repertoires of highly organized behaviors through observation, modeling, and learning situations. A person can select from potential abilities or known competencies that would be adaptive in the situation, based on his expectancy regarding the contingencies present in the situation, the relationship between the stimulus and outcome, and his subjective values.[5]

The choice of a specific behavior reflects in part the individual's expectancy regarding the likelihood of various outcomes contingent on the chosen behavior. *If* the person selects behavior X *then* outcome Y is expected. An entire series of if-then choices may be reviewed before the final choice of behavior is made. For example, when a person is asked for a date there are multiple if-then contingencies possible. *If* the invitation is accepted *then* no others for the same evening will be accepted, unless the contingency was included in the acceptance that if another, better invitation came then the first one would be cancelled. *If* the invitation is not accepted, *then* the person will not be busy that evening. *If* the invitation is accepted, *then* perhaps someone really interesting will be at the party. These are only a few of the if-then contingencies that may be present when a person is asked out. The alternatives available include acceptance, refusal, tentative acceptance, or a delayed decision. In the absence of new information concerning the various probabilities of the expectancies in any situation, the individual's choice will depend on past experience. If the probability is that this individual is never asked by more than one person for a date, then the expectancy will be that if the date is refused, the evening will be spent at home. The essence of adaptive performance is the recognition and appreciation of the significant contingencies present in a situation. The maladaptive person is behaving in accord with expectancies that do not adequately represent the actual outcomes present in the current situation.

The individual also learns that the presence of certain cues or stimuli in a situation are predictive of certain events. These cues are discriminative in nature and assist the individual in selecting the alternative that is most adaptive. A person's response to a physical blow from another individual will depend upon whether it is perceived as an intentional or accidental encounter. If it is perceived as intentional the person can select an alternative that will enable him to express displeasure, either physically or verbally; if, however, the perception is that the blow was accidental then the class of alternatives will be directed to reassuring the other individual that no harm was done, and that the lack of intention was noted. If the person is unable to discriminate the two stimulus situations the alternatives selected for a response often may be inappropriate.

The choice is also influenced by the subjective value an individual places on the various alternatives available for a response. Personal preferences and aversions will vary from individual to individual. When a highly preferred alternative is paired with an alternative that is avoided if possible, an individual will not have difficulty making a choice. The difficulty is present when an individual must choose between two highly preferred alternatives or between two nonpreferred alternatives. In these two situations the behavior will demonstrate ambivalence and instability until the choice is made. For example, most people have difficulty making a choice when confronted with a tray of French pastry just as most would have difficulty selecting a weapon for a duel.

The environment and specific situation serve to limit the available choices through both the perseveratory and preparatory sets. Certain choices of behavior that might be available in the home are not available in other settings. Age and health are important factors that serve to limit the range of available choices. A young child and an aged individual have a narrower range of choices available for selection than does an adolescent or middle-aged adult. The child's restricted range is due to a lack of experience with a variety of situations, plus a relatively restricted learning history; the aged individual is restricted by declining physical and mental abilities. No matter how limited the choices may be, however, a choice is always possible. In the case of a robbery victim who is told "your money or your life" the victim still has a choice, albeit between two difficult alternatives. The predetermined choice, present in situations where only one response is available, still contains the alternative of not responding, even though it might not be in the best interests of the integrity of the system to do so.

Action

The fourth structural component of the behavioral system involves the directly observable *actions* of the individual. Actions are dichotomous in the sense that they either occur or do not. If they fail to occur they are not available to the observer as a source of information concerning the current level of activity within the system. Based upon the observed actions a tentative explanation is made by the observer as to the intention of the behavior. Some actions appear to be directly goal related, such as an individual who is observed eating lunch by himself in the cafeteria. Other actions may have no immediately apparent purpose attached to them. If an individual is observed to be walking down the hospital corridor, the only statement that can be made reliably is that "she is walking down the corridor." Beyond that, an explanation of the purpose of the behavior can only be inferred. The inference may be based on prior knowledge of the individual, or on knowledge concerning the reasons most people walk down hospital corridors. At this level the observer is dealing in generalizations based on knowledge other than that contained within the immediate situation. A higher level of inference would attempt to form a theory regarding the multiple factors contained within the present situation and other similar situations that the individual of interest may be engaged in throughout a lifetime. The higher levels of inference are less directly related to the behavior observed in the immediate situation. Instead they are attempts to relate the observed actions to other knowledge concerning the individual. In a sense the highest level of inference constitutes hypotheses which can be subjected to test, either through inquiry or further observations.

A series of actions may have multiple goals. A man and wife attending a political dinner for which they paid $100 a plate may simultaneously be fulfilling achievement, affiliative, and ingestive goals. A patient who persistently rings the callbell for a nurse may be fulfilling dependency, affiliative, achievement, or even aggressive-protective goals. Unless an action is situationally defined, an observer will have difficulty in judging what the behavior means and how best to respond to it. Any tentative explanation as to the underlying meaning of a behavior must be tested by questioning the individual about the behavior and its intention, or by making repetitive observations of the same behavior to determine whether there are particular situations in which the behavior appears with the highest probability, as well as to determine those situations in which there is a low probability that the actions will be observed. Suppose that a female patient wears an extremely provocative nightgown only at certain times of the day. Upon questioning, her response is that there is no particular reason, but that she merely had nothing else to wear. Repetitive observations of the occurrence of this behavior and particular attention to the conditions or situations under which there is the greatest likelihood that she will wear this specific gown may result in the conclusion that it is related to expected visits from her fiancé since she consistently wears the gown when he visits, and rarely, if ever, wears it when other visitors come. The interpretation would be made that this gown is a choice which has been selected to fulfill the goal associated with the sexual subsystem rather than the affiliative subsystem, because of its direct association with visits from the fiancé.

Sustenal Imperatives

The fifth structural component of the behavioral system involves the *sustenal imperatives*, which consist of the necessary prerequisites for survival of the system. For a behavior to be maintained it must be protected, nurtured, and stimulated. *Protection* from

noxious stimuli that threaten the survival of the behavior is imperative; *nurturance* provides an adequate input to sustain the behavior;
stimulation contributes to continued growth of the behavior and
counteracts potential stagnation. A deficiency in any or all of these
sustental imperatives threatens the life of the system as a whole, as
well as the effective function of the particular subsystem which is
directly involved. Eating behavior is stimulated through hunger
pains, pleasurable aromas, the time of day, or even conversation
about foods. These same behaviors are protected by eating foods that
are neither too hot nor too cold, by avoiding poisonous substances,
and indirectly by enforcing standards established by the Food and
Drug Administration. Nurturance of eating behaviors is assured by
the immediate availability of foods, sufficient economic sources,
preparation of food by others, and encouragement to feed oneself.
It should be noted that many sustenal imperatives are provided by
individuals and agencies other than the immediately involved individual. A major function of the nurse can be conceptualized as protecting, nurturing, and stimulating patient behaviors that will
ensure survival of the system as a whole and that will maximize the
desired good for the individual.

These five structural components—goal, set, choice, action, and
sustenal imperatives—interact with one another to create the overall
function of the system. If there is a discrepancy in any one part such
that the component does not function within normal limits, the
function of the system as a whole will be impaired. The discrepancy
will often relate to an observed action having little or no relationship to the intended goal of the behavior. Many forms of mental
illness demonstrate this type of discrepancy between goals and actions in which repetitive behavior is indiscriminately employed in
a wide variety of unrelated situations. A second common discrepancy is a failure to discriminate significant factors in the immediate
situations which would preclude the use of previously learned responses, the perseveratory set having become rigid.

BEHAVIORAL SYSTEM FUNCTION

The behavioral system has been conceptualized as an integrative
response system which adaptively relates to various stimuli and com-
municates the status of internal processes to the surrounding environment. Therefore, behavior must have meaning to an observer.

If the message has no meaning, an observer arbitrarily assumes that there is "something wrong" and that the individual is unable to function adaptively in relationship to the demands of the environment. It is possible that the lack of meaning resides within the observer, who distorts or blocks the meaning of the message. However, it is also likely that the structural components of the behavioral system are dysfunctional in some way, such that the action of the individual is *maladaptive,* i.e., does not serve an adaptive function. The problem may be at the level of the goal, perseveratory set, preparatory set, or selected choice of response, and it is necessary to carefully evaluate the possible level of dysfunction in order to effectively intervene. It may be that the actions have been deliberately distorted by the individual in order to mask the true state of affairs, e.g., when an individual attempts to be polite to someone with whom he is extremely angry.

Every society provides rules to be used by its members for establishing meaning within a situational context. In this way any individual within the group will have a shared collective reference point for intepreting the possible messages being communicated by the behavior. Suppose an individual is observed in a cafeteria drinking coffee and engaged in conversation with the other people at the table. Three *manifest,* or distinct, messages are being transmitted to the observer:

1. the individual is ingesting coffee—an ingestive behavioral pattern.
2. the individual is associating with others, possibly friends— an affiliative response pattern.
3. the person appears to be relaxing in the social atmosphere of a coffee break—a restorative behavioral pattern.

These are a few of the more obvious interpretations of the manifest behavior any American might make of the observed situation, since the coffee break is a standardized or ritualized set of behaviors that everyone "knows" about and that involves a standardized set of actions considered to be adaptive for the situation.

Analysis of the situation at a deeper level of inference would require additional information concerning the motivations of that person in the situation and the relationship to others. This further analysis would reveal the *latent* function of the behavior. The individual may be drinking black coffee because he is on a diet; he is on a diet because he desires to become more successful with the opposite sex. The latent function of the behavior therefore is directed to a less obvious, long-term goal related to the sexual subsystem; the manifest function of the behavior is to ingest black coffee instead of a higher-calorie drink in order to attain the goal of the ingestive subsystem which is to ingest fluids when needed or desired. In general the *latent function* of a behavioral response will be related to a long-term goal whereas the *manifest function* will be determined primarily by the interaction of the short-term goal and the immediate situational demands.

Behavioral systems analysis is restricted to dealing with the obvious manifest meanings communicated by the behavior within the situational context in which the behavior is observed. It attempts to avoid the many problems created by higher levels of inference or by assumptions regarding the latent function of behavior, by making repeated observations of the behavior in a variety of situations in order to search out the possible underlying message of the behavior. If the same behavior occurs in very dissimilar situations it may be extremely difficult to identify the common aspect that is eliciting the behavioral pattern. It can only be assumed to exist. The same process can be applied to the study of maladaptive behaviors in which the individual is responding to noncritical components of the environment and the behavior does not convey meaning to others. Behavior analysis can make use of subjective reporting to verify observations or identify possible other critical aspects that are contributing to the observed behavioral response. The subjective report, however, must be considered in relationship to the observations and must not be used as the exclusive form of knowledge concerning the individual.

Finally, behavior attempts to elicit a response from the environment that is also available to the observer for analysis. It is possible that the same behavior is evoked indiscriminately in a variety of situations not because of a condition in the environment, but because it will elicit a response from the environment that the system needs for continued survival. It must be remembered that consistency of response does not necessarily mean that the response is adaptive; in contrast, it may be extremely maladaptive because of the very frequency with which it is employed. The adult who is constantly seeking the help of others rather than attempting to function independently has developed a consistent, but maladaptive, pattern of dependency subsystem function. A major purpose of such consistent responses is to maintain or replenish energy that the system needs to continue to function, even though the function is not totally successful in terms of health.

BOUNDARY FUNCTION AND REGULATORS

The boundaries of the behavioral system are highly flexible and continuously adjust to the level of input received from the surrounding environment in terms of the ongoing level of activity within the behavioral system. The function of the system is regulated to a large extent by the input received from the physiological, psychological, social, and cultural systems that constitute a major part of the immediate surrounding environment. The input received from each of these regulatory systems differs in form and intensity and must be transformed by the behavioral system into stimuli on which to establish a basis for responding. Input from one of the four regulatory systems may take precedence at a given moment, and as such will be the dominant determinant of observed behavior. For example, behavior at a church service is primarily regulated by specific cultural system input, whereas behavior at a dinner dance is predominantly regulated by social factors.

The boundaries of the system limit the amount of input made available to the system at a specific time. If the system is in danger of experiencing overload and the consequent disorganization of behavioral function then the boundary will tighten and exclude all but the critical input required for continued function. The function of the boundary is similar to that of sensory thresholds; in sensory-poor environments the input threshold will be lower than in complex sensory-rich environments. If the system is relatively quiescent, with a low level of activity, the boundary may allow stimuli of weak intensity or relatively low importance to enter. The ability of the boundary to function flexibly is of great importance to survival of the system. If there were no means of regulating the amount or intensity of stimuli allowed into the system for integration and response the system would rapidly disintegrate. Boundary function directly relates to the perseveratory and preparatory sets of the behavioral system, and its level of activity can in part be ascertained through analysis of these two structural components.

EQUILIBRIUM AND BEHAVIORAL STABILITY

Distribution of available energy among the eight subsystems is determined by whichever subsystem happens to be dominant at that
time. Each subsystem has a minimal requirement for energy to sustain basic activity. Beyond that, energy is distributed where it is
needed in terms of the ongoing response pattern, the situational
demands, and the input and feedback from the environment. The
usual interaction between subsystems results in an ever-changing
picture of energy distribution. Figure 2 illustrates a hypothetical
case in which the affiliative subsystem of behavior has become insufficient, or atrophied, for lack of stimulation. The energy normally

Figure 2
Behavioral System Disturbance:
Insufficiency of One Subsystem

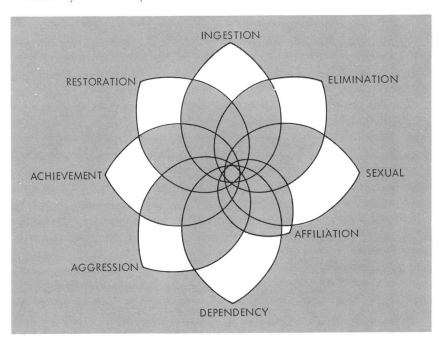

received by this subsystem has been redistributed among the remaining seven subsystems. This situation might be characteristic of the recluse or of any individual who does not develop affiliative behaviors to the degree required for survival. This is not meant to imply that all persons must be constantly seeking affiliative experiences, for some individuals require fewer contacts for their idiosyncratic level of function than do others.

Figure 3 illustrates the situation in which the affiliative subsystem has attained a dominant position and is a major response system for the behaviors engaged in by the individual. This results in less available energy for distribution to the remaining seven subsystems. If a condition of imbalance is sustained for an extended period of time, the ability of the system to respond efficiently to increased demands for energy is diminished. The pattern illustrated by Figure 2 would be apparent through an absence of or deficiency in the number of actions observed that are associated with the insufficent subsystem. The situation shown by Figure 3, dominance of one sub-

Figure 3
Behavioral System Disturbance:
Dominance of One Subsystem

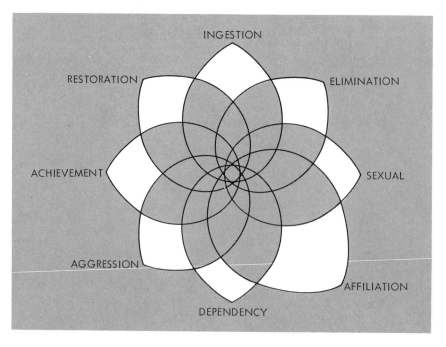

system to the exclusion of others, would be apparent through a preponderance of actions associated with that subsystem, and a tendency for infrequent use of response modes associated with the other subsystems.

Behavior, under nonstress conditions, tends to be organized and patterned in such a way that the individual can respond with a minimum of energy expenditure in a *stable* or consistent manner. Responses tend to be flexible and appropriate to the situation, reflecting a dynamic process of adaptation and balance among the various subsystems. Physical or social stress serves to disrupt ongoing activity and produces disorganization or *instability* of ongoing behavioral response patterns. Unstable behavior will be characterized by hesitancy, ambivalence, and fluctuation between various actions until an effective response is determined. The familiarity of the situation and past experience with the contingencies associated with the stressor will influence the degree of instability experienced by the individual. During such stress periods the system is in disequilibrium, and increased amounts of energy will be consumed in an effort to restore balance.

SUMMARY. This chapter has presented basic concepts of behavior in a systems analysis framework. Behavior is assumed to be a quality of an individual which can be treated separately from all other qualities. The structural parts of the system are represented by eight subsystems, each of which has a specific goal, set, choice, action, and sustenal imperative. The eight subsystems are:

1. Ingestive
2. Eliminative
3. Dependency
4. Sexual
5. Affiliative
6. Aggressive-protective
7. Achievement
8. Restorative.

The immediate input to the behavioral system are the biological, psychological, social, and cultural factors elicited by a situation. The overall goal of the behavioral system is to produce an integrated response to the surrounding environment in a form that has meaning and purpose to both the environment and the individual. The biological, psychological, social, and cultural systems represent major regulators of ongoing behavioral responses. For example, when an individual has the flu, the physiological system responds by developing a headache, sore throat, stuffed-up nose, bleary eyes, and muscular aches and pains. The psychological system responds by turning inward on the feelings experienced within and on memories of past experiences with the flu, rather than by turning outward to interact with others. The social system interrupts human interactions and isolates itself as much as possible, since it is not acceptable to infect others. The cultural system focuses on going to bed, drinking plenty of fluids, and being cared for. All these data constitute the immediate input received by the behavioral system. This input is then transformed and integrated into increased dependency and restorative behaviors; limited affiliative, achievement, and sexual behaviors; and modification in the specific direction of the ingestive, eliminative, and aggressive-protective behaviors during the illness. The resultant behavioral response pattern signals to an observer that the person feels ill.

Chapter 4 will describe the structure and function of the eight subsystems. Following chapters will discuss general regulatory factors of behavior.

STUDY QUESTIONS

1. Observe and analyze the behavior of an individual in terms of the specific situation and the manifest behavior.
2. What are some of the determinants of energy distribution within the behavioral system?
3. Analyze the pattern of behavior of an individual in the following situations, including the major regulatory factors.
   a. Classroom lecture.
   b. Initial behavior of a patient following admission to the hospital.
   c. Mealtime, alone and with a group.
4. What response patterns of behavior are available to the infant in the first week following birth?

four | *The Structure and Function of Behavior*

Behavior has been defined as those complex, overt actions or responses to a variety of stimuli in the environment that are purposeful and functional to the individual. The overall purpose of behavior is to maximize a potential or desired outcome for the individual or group. It may involve the acquisition of environmental objects or resources, or the avoidance of objects and situations that could result in the potential injury or death of the individual.

Specific behaviors are mastered by the individual throughout his lifetime and they acquire meaning and function in relationship to attaining the goal of maximum good. This does not imply that all behavior will appear meaningful and functional to the environment in which it occurs, but rather that the individual, based upon his perception and interpretation of the environment, has selected a pattern of response that has meaning and purpose to him. As long as the environment assesses the observed behaviors as meaningful and functional to the situation the behavior is termed *adaptive*. When meaning and purpose are not conveyed to the surroundings, the behavior will be viewed as *maladaptive* because of the lack of apparent relationship to events transpiring in the environment. A second criterion of the adaptiveness of a behavioral pattern is associated with the observed intensity of the response in relationship to the environmental conditions. Adaptive responses are characterized as appropriate in intensity to the eliciting stimuli, whereas maladaptive responses may be a deficient or excessive response to the event. Behavior is also characterized by a patterned organization which is relatively *stable* or consistent. *Instability* or disorganization will be observed when the individual encounters a physical or psychosocial stressor, or an unknown situation for which the individual has no immediately available response.

Eight subsystems of behavior have been identified by Johnson as representative of the range of human behavior.[1] Each subsystem is comprised of "a set of behavioral responses or responsive tendencies which seem to share a common drive or goal."[2] The perception of the individual and the integrated input from the biological, psychological, social, and cultural regulators serve the immediate function of eliciting a particular pattern of responsive tendencies in order to meet the goals of the individual. Since perception of a situation is a highly idiosyncratic process it is possible that different sets of responsive behaviors will be elicited by the same set of environmental conditions. For this reason the analysis of observed behavior can deal only with the apparent, or manifest intent of the behavior in relationship to the environmental events. Only if the psychological perception of the individual is available through self-report can the latent function of the behavior also be explored.

The following discussion of the eight subsystems of behavior focuses on the shared characteristics of behavioral actions within each of the subsystems. The division of behaviors into eight separate groupings is arbitrarily based upon an apparent relationship among various behavioral actions and their assumed function in an individual. It is possible that there are further differentiations of behavioral acts that have yet to be identified. It is also possible that these specific behaviors could be grouped in different ways to form a different set of subsystems. However it is assumed that the proposed differentiation on the basis of apparent shared characteristics of function in accomplishing a common goal is a useful and reasonable basis of categorization. Each of the subsystems will be described in terms of specific structural units and their function in the behavior of an individual. Various indications of malfunction within a subsystem will also be discussed.

INGESTIVE SUBSYSTEM OF BEHAVIOR

Structural Components

Goal: to internalize the external environment in order to maintain the integrity of the individual and/or to achieve a state of pleasure. Specific goals include: internalization of food and/or fluids; acquisition of information to relieve a state of ignorance; acquisition of resources required for continued survival, such as oxygen; monitoring of environmental events for changes and resources that constitute the input for activation of the remaining seven subsystems.

Set: an awareness of a state of deficiency existent within the individual that can be corrected through ingestion of environmental resources. The perseveratory set involves the predisposition to discriminate specific events or resources in the environment in a habitual pattern.

Choice: the range of alternatives available to meet an ingestive goal is extensive and includes: formal education; self-study; availability of a variety of preferred food and fluids in the environment; restriction of intake; exploration of new potential resources; curiosity.

Acts: any directly observable behavior that serves the purpose of increasing intake of environmental resources related to the condition or felt deficiency. Specific acts include eating, drinking, reading, listening, seeing, touching, inhaling, and other actions that serve the purpose of taking in conditions and resources existing within the surrounding environment.

Sustenal imperatives: conditions that serve to protect, stimulate, and nurture behaviors related to ingestion. Included are the learned associations of effective ingestive actions and the reduction of felt deficiency; avoidance of harmful and poisonous food and fluids; enriched stimulus environment; physical settings that contain adequate supplies of required substances; conscious awareness of deficiency states; prior success in the reduction of deficiency; economic resources; and awareness of existing resources.

Function of Ingestive Behaviors

The general function of the ingestive subsystem of behavior can be described as the means by which an individual is brought into relationship with his surrounding environment. This relationship is essential for the identification and ingestion of those energy supplies required for survival. Sensory information enables the individual to initiate required compensatory responses in order to maintain a state of psychophysiological equilibrium with changes in the external environment. Further, it is through the ingestive activity that the individual is able to differentiate himself as a separate entity from all other objects and persons existing in the outer world.

The process of perception provides the individual with large amounts of information, all of which must be interpreted in terms of the prior and current status of the individual. A malfunction or absence of one or more sensory modalities may severely handicap the relationship of the individual to his environment. The absence of one sense modality can usually be compensated for by increased sensitivity or environmental monitoring by another sensory channel. For example, blind people are often more attentive and sensitive to auditory information than are sighted persons. Some learn to judge the distance of physical objects by the quality of sound reflected by the objects, which is the identical process used by dolphins, for example, to locate food sources in the depths of the ocean. A person who is both blind and deaf is severely handicapped in terms of establishing an ingestive relationship to any but their immediate environment and, as a consequence, the ability of the environment to provide meaningful input is blocked. The entire pool of environmental stimuli and resources available to these individuals becomes limited to that which can be sensed through the existing senses of taste, smell, touch, pressure, temperature, and pain. Consequently the movement of the individual throughout the extended surroundings is limited. A child who is deaf and blind from birth has an extremely difficult time developing a repertoire of stable ingestive actions within the environment. One child had to be placed in a cardboard box which provided a known limit to the seemingly endless abyss in which she existed. Once she had experienced this structured boundary, her behavior became patterned and organized, i.e., stable, and she was able to gradually extend this boundary without returning to her prior state of disorganization.

The importance of this behavioral response system for the physiological survival of the individual is paramount. If physiological survival is not ensured, the sociocultural and psychological aspects of the individual become superfluous. Persons who were interned in concentration camps during World War II were often unable to think or fantasize on any subject other than food during severe starvation periods. Much the same thing occurs when an individual first begins a diet. The topic of food and the desire for foods that have been banned become the dominant concern of the individual. If the ingestive subsystem is prevented from functioning in terms of locating the required energy supplies in the external environment to correct the experienced state of deficiency, the entire individual as a system becomes threatened by extinction.

The ingestive subsystem assumes an intimate relationship to all other subsystems through its function of internalization of external stimuli. The information acquired through the perceptual process forms the basis of the preparatory set of the other subsystems and assists in identifying possible choices that exist within the setting. For example, the ingestive subsystem function provides the knowledge to the individual who requires assistance (dependency goal) that certain individuals exist in the situation who can provide help. Absence of this information would severely limit the function of the dependency subsystem and the responsiveness of the behavioral system as a whole.

Malfunction of the ingestive subsystem may be characterized as hyper- or hypoingestion of the environment, and distortion of perceptual contents. Anorexia nervosa, a disturbance characterized by severe restriction of food intake, represents an extreme form of *hypoingestive* behavior that can result in death if more adaptive behaviors are not nurtured and stimulated in some form. This disturance most commonly affects girls entering adolescence and is hypothesized to be an unconscious attempt to avoid growing up and the threatened loss of dependent relationships with others. Treatment of this condition requires stimulation of more adaptive ingestive behaviors and nurturance of adaptive dependency and affiliative behaviors that enable the individual to establish a mature relationship with others. The behavioral patterns associated with ingestion for patients with anorexia nervosa are quite different from those patterns associated with environmentally induced starvation. In the latter case the individual may demonstrate a dominance of ingestive behaviors related to attempts to constantly monitor the environment for any potential food source. In this sense these individuals may be described as exhibiting a *hyperingestion* of the environment. Obesity also represents a pattern of hyperfunction of the ingestive subsystem in which innate limits of satiation do not serve to interrupt the behavior when energy requirements have been assured.

Distortion of perceptual contents can be observed through disorganized or apparently meaningless behavior associated with other subsystems. For example, paranoid thinking can be conceptualized as a distortion of the environment in which various aspects are interpreted to be threatening to the individual. The distortion will be apparent through activity associated with the aggressive-protective subsystem in situations that do not contain any identifiable threat to the person. Distortion of perception may also occur when the environment contains too many or too few sources of stimuli. The situation of sensory overload places demands upon the ingestive subsystem which may exceed the normal limits of informational processing. Unless the boundary can become less permeable, behavior will become disorganized and fragmented. Sensory deprivation, on the other hand, creates a condition in which the ingestive subsystem does not receive sufficient stimulation from the environment and the internal processes attempt to "fill in the gaps," resulting in perceptual distortion.

ELIMINATIVE SUBSYSTEM OF BEHAVIOR

Structural Components

Goal: to reduce a perceived state of tension or pressure existing within the individual in order to maintain the integrity of the individual and/or achieve a state of pleasure. Specific goals include: elimination of body wastes through excretion or rejection of surplus; nonverbal and verbal behaviors intended to communicate to the external environment the expression of feelings and emotional states contained within; the imparting of information.

Set: an awareness of a state of tension which can be relieved through a release of energy to the environment, or the desire to impart knowledge for the pleasure created within through the act of sharing; perception that a situation exists in the environment which requires information that is contained within.

Choice: the basic choice available to the individual is to hold in or let go. Unless release of tension in some form occurs voluntarily, buildup of contents will eventually result in the involuntary release of contents.

Acts: any directly observable behavior which serves the purpose of tension reduction for the individual. Specific behaviors include: urination; defecation; perspiration; menstruation; exhaling; speech; nonverbal gestures; expression of affective states through such behaviors as crying, yelling, laughing, smiling, smoking, doodling; and other actions that serve to communicate the general internal state of the individual to the environment.

Sustenal imperatives: conditions that serve to protect, stimulate, and nurture behaviors related to elimination. Included are the learned associations of environmental conditions that facilitate expression of internal states as distinguished from those that do not favor immediate expression; establishment of a supportive, accepting environment; ingestion of foods and fluids; physical exercise; development of autonomy; availability of physical facilities; prior success in reduction of tension states; and awareness of the amount of pressure that can be tolerated by the individual prior to a loss of control.

Function of Eliminative Behaviors

The general function of eliminative behaviors is the externalization of the status of the internal environment. Whereas the ingestive subsystem serves to monitor and incorporate the external surroundings as an informational process for the remaining seven subsystems, the eliminative subsystem transmits information to the environment as to the status of the system as a whole. The ability of man to recall events and store memories enables transmission of information that relates to the state of the individual in the past· as well as permitting comparison of past conditions with the current state.

It is primarily through the eliminative behaviors that a second individual—the observer—can ingest knowledge about the state of the first individual. Suppose that the internal experience of a patient is one of anger and hostility. There are a number of alternatives available to the person to externalize this feeling state and possibly modify the environment in order to reduce the intensity of the affect. One of the most obvious possibilities is to verbalize the state by saying, "I am angry." At the same time other forms of eliminative behaviors would assist in the assessment of whether or not the release of information is a valid indicator of the internal state of the individual. Concurrent with the verbal message the following observations might be made: flushed face, tensed muscles, narrowed eyes, clenched fists, and a mouth clamped tightly following the declaration. All these nonverbal eliminative behaviors combine to form an integrated pattern that enables the observer, as the external environment, to tentatively conclude that the individual is experiencing a state of internal tension that can be labeled anger. If the individual made the same statement but was laughing or smiling, an ambiguous message would be communicated that would require additional information before the verbal or nonverbal content could be validated. Nothing beyond the knowledge that a particular state of anger exists is known, however. One does not know the situation or condition which has produced this state of tension, nor whether it was elicited by the external environment or was a consequence of stimuli contained within the individual.

A second example of the informational function of eliminative behaviors can be related to the routine examination of urine. The urinalysis represents the use of a physiological eliminative waste-product to evaluate the biological system and to provide information concerning the fluid balance, electrolyte balance, kidney function, nutritional status, endocrine metabolism, and so forth. Essentially it provides general, nonspecific information that relates to many aspects of the individual. It is important information that may not be available to the individual on any but the biological level.

How valid is the eliminative information that is made available by the system? Essentially, the problem is a matter of determining the degree of validity between the actual internal state and that which is externalized by the individual as the *apparent* state. The answer must be that there is always the potential of error contained in the information. Some of the error may be accidental; perhaps the patient forgot to obtain a clean specimen for a urine culture and the result of the test indicates that a large number of bacteria are present. Under such conditions the individual is erroneously communicating information which results in a conclusion that a condition of illness exists within the individual.

It is possible, however, that the error may be purposeful on the part of the individual. To a large extent the verbal communications used to provide information to the external environment concerning internal thoughts and feelings are under the conscious control of the individual, and have been monitored in terms of sociocultural and interpersonal acceptance prior to externalization. Only that information that the system desires to externalize will be made available; the remainder will be contained within the individual. Furthermore, the individual may distort the information so that the true state remains unknown. Distortion may be voluntary or involuntary; the individual may not be aware that he is misinforming the environment because he is unaware of the conditions existing within himself.

Certain patterns of eliminative behavioral actions are more open to error than are others. Biological factors are, to a large extent, not subject to the voluntary control of the organism and are therefore less subject to conscious distortion. Much of the popularity of biofeedback approaches as adjuncts to psychotherapy is based on the valuable information gained through monitoring changes in alpha rhythms, finger temperature, skin resistance, and blood pressure as they relate to verbal content. It was on the basis that biological factors are less subject to voluntary control that the lie detector test was developed, on the assumption that responses of organs innervated by the autonomic nervous system are not under conscious control of the individual. Therefore, if a question is asked of the individual and the verbal response is an attempt to intentionally deceive, or to distort the truth, this deception will be manifested by changes in the biological indicators of heart rate, respiration rate, and electrical skin resistance in the directions that are congruent with a state of arousal.

The verbal channel of eliminative behaviors may also be subject to involuntary distortion. Freud attempted to analyze "slips of the tongue" and their possible relationship to internal thoughts and feeling states.[3] He felt that these slips, or accidents, were representative of a loosening of the controls which prevent the true expression of internal conditions, and that the content expressed by the slips was of greater significance than the content that was being consciously verbalized. Jung explored the nature of the relationship between the presentation of word lists, termed *stimulus words*, and the response associations of a given individual.[4] He examined the quality of the response to determine whether the association was a common one such as book-page, father-mother, teacher-student, in contrast to less apparent associations such as book-look, father-locket, or teacher-tree. He found that if the stimulus word was related to a highly charged emotional area of significance to the person that the response was characterized by hesitation, blocking, mistakes, stammering, or there was a prolonged period of time before a response could be given. Jung felt that the externalization of areas of unconscious conflict, or *complexes*, could be identified through administration of such word-association tests. Disturbances in psychic processes are often reflected through maladaptive eliminative behaviors such as loosening of associations, flattened affect, neologisms, and other changes in the normal stream of thought.

Malfunction of the eliminative subsystem represents failure to externalize internal contents, i.e., denial and repression; excessive externalization of internal contents; and inaccurate or distorted externalization of contents. Failure to directly release the internal tensions may have serious consequences to the individual, ranging from loss of voluntary control over micturition and defecation, to the possible development of a psychosomatic disease. Theories of psychosomatic illnesses postulate that failure to express affective states through such behavioral actions as crying, yelling, laughing, and direct verbalization will result in eventual symptom formation that indirectly reflects the underlying pathology.[5, 6] A second consequence of the failure to release or eliminate internal pressures is the potential loss of voluntary control over the process once it is initiated. Many people have attempted to hold in an intense feeling, only to have it break through and threaten to completely overwhelm them and the object toward whom the affect is directed.

Excessive externalization of internal states results in a never-ending stream of thoughts and feelings with little evaluation as to the meaning or content. Such individuals are often described as people who "like to hear themselves talk," or in the terminology of the eliminative subsystem, these people have verbal diarrhea. Inaccurate or distorted output may be accidental or purposeful. Incongruency between the content of information conveyed over the verbal and nonverbal channels is a common manifestation of malfunction. In the example of the individual who was expressing the verbal statement of anger while nonverbally communicating actions that are usually associated with feelings of happiness and joy, the two messages were incongruent. The observer must then determine the source of the incongruence and help the individual develop a congruent mode of communication. Many clinical psychiatric disturbances are manifested through malfunction of the eliminative subsystem of behavior. For example, thought disturbances characteristic of schizophrenic processes, including loosened associations, concreteness, flattened affect, flight of ideas, and autistic actions based upon hallucinatory processes are all expressed through eliminative behaviors.

SEXUAL SUBSYSTEM OF BEHAVIOR

Structural Components

Goal: to procreate and/or obtain gratification from sexual activities. Specific goals include: development of an appropriate identity based upon one's sex; development of skills appropriate to one's sex; assumption of societal status and roles prescribed for one's sex; and establishment of physical intimacy.

Set: awareness of one's biological sex and the behaviors expected of persons of that sex; development of a set of specific behaviors appropriate to heterosexual interactions that are distinguished from the set for use in same-sex interactions. The perseveratory set involves the predisposition to relate to others in a consistent pattern of sex-linked behaviors in an extensive variety of situations.

Choice: the range of alternatives provided for establishment and maintenance of sexual behaviors is extensive, but is limited by the expectation that individuals will select social roles, objects, and physical intimacy with others that are congruent with their biological sexual identity. Therefore, the choices are basically differentiated from birth. Within the general limitations, however, the individual may select from a wide variety of alternatives related to occupation, dress, leisure activities, and methods of attracting the interest of individuals of the opposite sex.

Acts: any directly observable behavior that serves the purpose of fulfilling one's sexuality, and which promotes survival of the collective society and culture. Specific acts include: pregnancy; intercourse; dating; housework; sex-linked clothing; and other actions that serve to identify oneself with persons of like sex. Also included are affiliations with groups based upon a sexual identity such as the Playboy Club and women's consciousness-raising groups.

Sustenal imperatives: conditions that serve to protect, stimulate, and nurture behaviors related to sexuality. Included are social rituals such as marriage; incest taboos; personal hygiene; awareness of one's own sexuality; pleasure seeking; perfume; after-shave lotion; birth; sex education; secretion of gonadal hormones; successful sexual interactions; physical attractiveness; plastic surgery; and physical environments that encourage expression of sex-linked behaviors, such as cocktail lounges.

Function of Sexual Behaviors

The general function of the sexual subsystem of behavior is to establish and fulfill the environmental expectations associated with one's biological sex. These expectations are socioculturally based and serve the function of ensuring cultural survival through the establishment of future generations.

The behaviors associated with the sexual subsystem exert a major influence upon the total function of the individual from the moment of birth. In the United States, individuals are reared from birth in the sex that is congruent with the biological characteristics present at birth. Cross-cultural studies report that males and females tend to differ from one another in a fairly universal manner, resulting in the conclusion that "despite the malleability of the infant organism, most cultures have chosen to satisfy their basic functions by capitalizing on these differences at birth rather than by ignoring them."[7]

The term *sexual* is not restricted to its narrow meaning of sexual intercourse. Behaviors related to this area include the complex role expectations associated with kinship relationships, such as husband, wife, father, mother, son, and daughter; the external accompaniments to behavioral acts such as dress and makeup; and the development of a sexual identification that is congruent with one's biological sex. Essentially the term sexual is used to connote those behaviors that are sex-linked and serve to differentiate male from female. This differentiation does not imply a value judgment in which behaviors associated with one sex are "better" than those commonly associated with the other sex. These value judgments reflect cultural values that may no longer be appropriate to the social structure of the group. For example, an agrarian culture will develop values for sex-linked behaviors that are different from a technological culture in which activities and expectations are different. Those cultures that have undergone a transition from one form to another, e.g. Japan and the United States, have concomitantly experienced a period of cultural instability during which sex-linked behaviors, as well as those behaviors associated with other roles, have been examined for relevancy to the new cultural structure. The biological differences between the two sexes are fairly self-evident; that males and females might be differentiated by such criteria as the ability to think rationally or the ability to cook is a function of sociocultural values and customs.

Development of a sexual identity emerges out of the overall formation of the self-concept of an individual. The environment provides the necessary information of how one is similar to, and different from, all others that exist. The label *boy* or *girl* is applied as a criterion of difference from the moment of birth and from that time on continuously influences the course of behavioral development. Parents tend to reward behaviors that are congruent with the labeled sex of the child and to punish those behaviors that are valued for the opposite sex. Manipulation of rewards and punishments produces sex-appropriate behaviors at a very early age. By the age of two it is possible to distinguish boy-girl modes of aggression and play, for example, and performance of physically aggressive actions by boys will be rewarded with approval, whereas the same behavior by a girl will be met with disapproval and punishment. Society expects that the sexual label based on biological factors will be congruent with the self-perceived identity, which is called the *gender identity*. That these two forms of sexual identification need not be congruent is demonstrated through the study of individuals who, although biologically classified with one sexual group, develop the psychological and social orientation that is consonant with the opposite sex.

Establishing a relationship with another individual that involves physical intimacy is a function of the sexual subsystem. This form of relationship may be a direct result of an intense, interpersonal affiliative relationship in which sexuality is used to express a love or liking for another individual. The only acceptable justification within Western cultures for failure to establish physical intimacy with another individual is a religious vow of celibacy; otherwise such failure is viewed as aberrant and indicative of a malfunction within the sexual subsystem.

The high level of sociocultural anxiety surrounding the area of fulfillment of sex-linked expectations is reflected by the extreme rigidity with which behavior is evaluated. Society, as a whole, rejects behaviors commonly associated with members of the opposite sex as evidence of malfunction. Penalties for actions such as homosexuality and transvestism, or, until recently, selection of an occupation that is identified with members of the opposite sex has included social ostracism and such labels as "sick," "abnormal," and "degenerate." Concern about sexuality and about whether one is fulfilling all that is associated with the ideals of masculinity and femininity is particularly prominent in the United States. Discussion of sexual problems with others is not encouraged and as a consequence many individuals suffer from a lack of knowledge concerning the adaptive function of this subsystem. The surest way to write a best seller is to write a book dealing with sexuality. *The Joy of Sex,*[8] *The Sensuous Woman,*[9] and *Everything You Always Wanted to Know about Sex, But Were Afraid to Ask*[10] have served the function of providing information in an area that has been characterized by a great deal of misinformation, although even these books are not totally accurate. Many individuals are afraid to expose their lack of knowledge to others and are more comfortable reading a book. This discomfort is one inheritance of the Victorian prohibitions against discussion of sex between male and female, and the comparison of the act of intercourse to behavior appropriate only to the lowest form of animal. The Victorian ethic still affects sociocultural values associated with sexual behavior, particularly among the middle-aged and older portion of the population.

Western cultures are now examining the values that have been associated with sex-linked behaviors, as to whether the expectations are appropriate or in fact discriminatory. During a period of cultural instability, behaviors that have been viewed as maladaptive to the subsystem may become accepted and valued, whereas those behavioral actions that were previously viewed as adaptive may no longer be considered to be positive. Before the advent of the women's liberation movement, employment of married women was considered appropriate only for those who had to work in order to provide the basic requirements for their family. Elective employment was viewed as evidence of malfunctional behavior. Now many married women with families are employed outside of the home because they enjoy the sense of fulfillment associated with employment. A variety of occupations that were formerly sex-linked are increasingly being opened to people regardless of sex, as the cultural values become flexible.

The American myth of sexuality used to be one of eternal youth in the sensuous form of Marilyn Monroe or Clark Gable. That few individuals ever attain this mythical apex is self-evident; that many judge themselves by such standards and find themselves short of the ideal is also true. Perhaps in the future the important functions of masculinity and femininity will be more clearly defined, and the rigid stereotypes of the past will be stripped away.

AFFILIATIVE SUBSYSTEM OF BEHAVIOR

Structural Components

Goal: to belong to or be associated with others in some form of a reciprocated relationship. Specific goals include: to be ascribed to a group without self action but with full membership status—family, class, religion; to achieve membership in a group through self-action—political party, professional and social groups.

Set: awareness of the need to be with others and of the particular role behaviors required by the specific setting of interaction. The perseveratory set involves the consistent tendency to select a certain individual or group for the purpose of affiliation to the exclusion of all others, and the consistent approach to the establishment of affiliative relationships.

Choice: the range of alternatives available to meet an affiliative goal is extensive and includes affiliation with members of ascribed relationships; development of substitute or supplemental achieved group memberships; avoidance of nonreciprocated relationships; affiliation with animals or other objects that provide the individual with a feeling of belonging.

Acts: any directly observable behavior that facilitates movement toward others in the environment. Specific acts include smiling, eye contact, social greetings, initiating conversations, use of first names, attending social functions, marriage, talking, corresponding, extending invitations, and other actions that serve to establish and maintain a reciprocal relationship between two or more individuals.

Sustenal imperatives: conditions that serve to protect, stimulate, and nurture behaviors related to affiliation. Included are the learned associations of prior successful affiliative relationships; development of trust; kinship; ritual kinship, e.g., adoption; presence of others in the environment with whom affiliation can develop; development of mutual interests; membership in a religion or religious institution; awareness of one's own individuality; self-esteem; presence of a variety of potential affiliations; ability to communicate; and knowledge of the formal and informal guidelines for the interpersonal process.

Functions of the Affiliative Behaviors

The general function of the affiliative subsystem of behavior is to establish a sense of relatedness and belonging with others. This desire to relate oneself to another is basic to all forms of human interaction. Although other subsystems of behavior contain an interpersonal component, each has a different function that underlies the specific purpose of the relationship. The interpersonal component of the sexual subsystem contributes to the establishment of a sexual identification through the parent-child interaction, as well as toward the establishment and maintenance of sex-linked relationships; however the relationship between two individuals is a means to an end, not the goal itself. Similar comments can be applied to dependency, aggressive-protective, and achievement behaviors also, as all contain a significant interpersonal component. The important distinction, however, is that an interaction between two or more persons is a function of the affiliative subsystem of behavior when the interaction is such that it provides pleasure and a sense of relatedness to *all* members of the relationship.

Every individual can relate to or associate with any number of individuals and groups throughout a lifetime. Two areas that distinguish individuals is the number of people with whom friendships and relationships are formed, and the depth of the relationships. People develop characteristic ways of relating to others. Some prefer to maintain relationships on a superficial level with most individuals, and develop a few, intense relationships with a select group.

Others are unable to develop a limited number of interactions of any sort. The rate at which an individual seeks out relationships with others is an important variable of subsystem function. One group of individuals can be described as high affiliators. These people tend to be socially extroverted; a large amount of energy is directed toward maintaining relationships in the interpersonal realm. They seek out interactions, are friendly and outgoing and thus tend to attract the interest of others. The low affiliator, in contrast, often has a low interaction rate with others and may depend instead on person-surrogates such as books, television, animals, and other objects to avoid a sense of loneliness and unrelatedness. Most individuals fall in between these two extremes, and use both people and person-surrogates to fulfill the goals of affiliation. Even when an individual is a "joiner" and acquires membership in many groups he would not be described as a high affiliator unless he actively participates in the groups. Others may not join many groups, but participate actively and frequently in those groups that are joined.

Intensity of interaction refers to the amount of effort the individual invests in a relationship and the emotional response that characterizes the interaction. Words such as love, like, tolerate, accept, reject, and hate are indicators of the intensity of the particular interaction and provide clues as to the desire for continued interactions in the future. If an individual dislikes another, the rate of interaction will decrease in proportion to that dislike, to the point of minimum requirements in relationship to the situation. For positive relationships in which pleasure is obtained, the tendency is to increase interaction to the upper possible limit. Rubin concluded that a reliable behavioral correlate of love was the frequency of eye contact between the two individuals, another indicator of the intensity of a relationship.[11] Pairs of individuals who were "in love" had higher frequencies of eye contact than did pairs of strangers.

The initial affiliative experience takes place within the family, for most persons. The emerging sense of identity and of selfness develops through interactions with persons who are similar and dissimilar. The family contains both types of experiences, for with the possible exception of identical twins, no two members of a family are similar in all dimensions, and it is through the experience of these differences and similarities that one comes to know oneself. For example, the daughter of a family will share the similarity of sex with mother and sisters, but may share similarities of interests with father and brothers. The sociocultural environment further formalizes these similarities and dissimilarities by expecting the individual to develop certain characteristics of behavior as a result of being assigned to various reference groups based upon race, age, sex, biological factors, nationality, class, status, and religion. Each of these sociocultural groups tends to restrict the range of possible behaviors allowed its members. Even though a person may voluntarily or involuntarily change membership groups from those of birth, he is still a product of the important early affiliative experiences that were influenced by specific reference-group memberships.

As the individual matures, a second major class of reference groups becomes available for membership. In these groups the person must make a choice as to which groups he desires to become associated with, and then earn membership in those groups. Whereas the family constitutes an ascribed group whereby the individual has little to do with becoming a member since it occurs by virtue of birth, achieved groups require the active participation of the aspiring candidate. Professional groups, such as the American Nurses Association or the American Bar Association, require that the candidate for membership have completed the educational requirements for practice of the profession. To become a member of a new religious group there are often requirements that the potential convert have studied and fulfilled various conditions prior to being accepted into the new congregation. These achieved groups reserve the right of evaluating the qualifications and interests of the potential members, and may reject the application of those individuals with whom they do not care to affiliate in a reciprocal relationship.

Malfunction of the affiliative subsystem of behavior is evidenced through failure to develop and maintain reciprocal interactions, and the associated sense of nonbelonging. Some people attempt to develop relationships but never seem to be successful in accomplishing their goals. Examination of their behavioral actions may indicate that whereas their manifest behaviors initially appear to be those that would encourage the development of a relationship, closer examination indicates that the behaviors are counterproductive. For example, suppose an individual attends a social function with the intention of meeting people, but every time someone speaks to him he turns away or responds with a hostile or biting comment. The goal for the individual would seem to be affiliative in nature, as the behavior is an attempt to contact and interact with others. However, the behavior has the consequence of pushing people away. It is possible that this person does sincerely want to make friends, but does not know how to proceed and has attempted to develop those behaviors which he thinks are positively valued by the group, but in fact are not.

Withdrawal into oneself may also indicate a malfunction of the affiliative subsystem, although it need not be the case under certain conditions. Withdrawal associated with the grieving process may initially be healthy and helpful to the recovery from a perceived loss of a relationship, either through death or separation by distance from a significant other. Extended withdrawal and isolation, however, cuts the individual off from the possibility of recovery through the development of substitute relationships with others. The loss becomes self-perpetuating, as the withdrawal from others removes the individual from assistance of others, until the individual has nothing but the grief to give meaning to life. There are times, however, when withdrawal from social activities may be quite healthy as persons evaluate themselves and their lives in order to attain a firmer sense of their own identity as separate from their relationships with others. In such situations of temporary withdrawal, the affiliative relationships are not severed, but temporarily suspended.

The mentally ill individual is often hospitalized when his behavior becomes so unpredictable or bizarre that his immediate family or friends can no longer understand or tolerate his presence. The unpredictable behavior usually involves the interpersonal process in which familiar clues of communication are absent. The individual may be attempting to establish contact through acting out behaviors directed toward himself, others, or property; he may withdraw and avoid contact; he may attempt to manipulate and control the situation and those persons in the situation by his very unpredictability; or he may develop a socially unacceptable form of relationship that constitutes a possible threat to life, such as drug addiction or alcoholism. A factor common to many forms of mental illness is that the individual has failed to develop a meaningful relationship with another person, and that he consequently feels an outcast and a failure. Such individuals appear to have no other mode of response available to them but negative affiliation with others, which only serves to perpetuate the situation. Severe malfunction of the affiliative subsystem of behavior, when it involves development of psychic disturbances and problems of relating to the sociocultural surroundings, may require professional help to assist the person in experiencing a genuine, healthy relationship that is pleasurable and valued by both parties.

Ongoing affiliative relationships are disrupted by critical events such as death or hospitalization. The individual who is cut off from persons who normally fulfill the requirements for an affiliative interaction may experience a period of instability while new relationships are being formed. During such periods the use of person-surrogate objects may become more important than usual. The individual may watch television more than usual, or seek consolation from religious practices that are normally not attended to under other circumstances. It is important to consider the impact of hospitalization on normal sources of affiliation for the individual, in order to sustain adaptive behaviors as much as possible.

DEPENDENCY SUBSYSTEM OF BEHAVIOR

Structural Components

Goal: to seek assistance from environmental resources in order to obtain a specific goal or assistance with a task. Specific goals include: seeking attention or recognition in a nonreciprocated interaction with another individual; gaining approval or reassurance from others; seeking assistance in a task from a designated source of authority.

Set: an awareness of an area of deficiency within the individual that can be corrected through the assistance of environmental resources. The perseveratory set involves the predisposition to select the same resource for assistance in all areas of perceived deficiency, as well as the predisposition to view oneself as helpless.

Choice: the range of alternatives available to meet a dependency goal is extensive, and includes: request of assistance from an individual who is an acknowledged expert in the area; avoidance of situations that relate to areas of deficiency; establishment of substitute objects that provide a sense of security or attention; exploration of environmental resources; and purposive exposure of oneself to situations that require assistance in order to gain increased self-reliance.

Acts: any directly observable behavior that serves to enlist the assistance of persons and/or objects in the environment that can reduce the experienced deficiency. Specific acts include: verbal requests for assistance; use of security objects; establishment of physical contact; cries for help such as suicide attempts; demands for attention through positive actions such as holding up one's hand in class or volunteering to assist authority figures, or through negative actions such as hitting peers or disrupting classroom situations; and other actions that serve to attract the attention and assistance of the environment.

Sustenal imperatives: conditions that serve to protect, stimulate, and nurture dependency behaviors. Included are the responsiveness of the environment to requests for assistance by the individual; the degree of knowledge of one's own strengths and limitations; prior success in enlisting the assistance of others; the degree of support, both task and emotional, available within the environment; and learned ability to obtain assistance and reassurance from others in an effective manner.

Function of Dependency Behaviors

The general function of the dependency subsystem of behavior is to obtain the assistance of persons and objects in the environment who are viewed by the individual as containing the knowledge and ability to assist the individual in meeting his goal in an area of felt deficiency. The deficiency may be a task-related lack of knowledge, either due to a lack of prior experience with the requirements of the task or through a lack of interest in developing the requisite skills required for self-reliance; or the deficiency may be related to the requirement of all individuals for continuous feedback from the environment as to whether desired goals are being accomplished, as well as a need for attention, reassurance, and approval of behavioral actions. For example, an actor in a play is dependent upon the audience to applaud his performance as a form of feedback regarding the quality of his performance.

Each individual interacts with and depends on many others within the environment to provide various forms of assistance. Infancy is an extended developmental phase of almost total dependency on others for the provision of the basic necessities of life. Part of the maturational process involves the discrimination of situations in which dependency on others is appropriate and approved, and those situations in which independence, or *inward dependency* is appropriate. Independence of function does not mean that an individual does not engage in activities related to the dependency subsystem of behavior. The complexity of our society requires interdependence among all of the units in order to sustain the individual members of the group. Self-reliance is an altered form of dependency behaviors learned through prior dependency interactions. For example, the self-reliant person, who is able to view his assets and liabilities without the direct assistance of another person has most likely developed this ability through prior interactions in which someone has provided the required feedback for the development of that self-knowledge.

The interactions that characterize a dependency relationship are intimately based on the affiliative subsystem regulators. The primary distinction between the behavioral actions associated with the two subsystems is that dependency behaviors do not involve reciprocity in the relationship. The dependent relationship is unidirectional between help seeker and help giver. In a pure dependency relationship, the help seeker will be the initiator of the interaction, such as in hiring a carpenter to perform tasks for which the individual has neither the ability nor the inclination. Many affiliative relationships are based upon alternating positions of dependency giver and dependency seeker. It is important to discriminate when such relationships are primarily directed toward the mutual pleasure gained from the interaction, or when the intent is to gain assistance.

The individual from whom help is sought generally is viewed as an expert within the area of felt deficiency. If help is needed with a specific task, assistance is usually obtained from individuals in the service-oriented occupations who identify themselves as expert within particular areas. The establishment of a person-dependency relationship in which the personality characteristics of the help giver are the crucial factors is much more complex and less readily apparent through direct observation. In general, the person toward whom dependent behaviors are directed will be one who appears to possess those personal characteristics that are identified as "good" or "desired" by the help seeker. The development of a successful dependency interaction, be it person or task oriented, requires the presence of two essential conditions:

1. the help seeker views the persons to whom he is directing his request for assistance as capable of providing the assistance;
2. the person who is requested to help is willing and able to fulfill the requested assistance.

If either element is missing, a dependency relationship will not be established.

Task-oriented dependency behaviors that are directed toward the completion of a task through the assistance of others are usually of short duration, and specific to the task. These behaviors are often used to attain an immediate short-term goal, whereas the long-term goal may be more closely related to another behavioral subsystem, such as achievement. For example, a nursing student may request assistance from the instructor when administering medications until he or she has sufficiently internalized the various components of the task. This request for assistance is related to the short-term goal of becoming proficient in giving medications, and to the long-term goal of becoming a nurse. As the individual develops an extensive repertoire of behavioral acts, fewer task-oriented dependency behaviors are observed, and the behaviors become related to highly specific, well-demarcated areas of deficiency.

Malfunction of the dependency subsystem can be classified as *excessive* or *deficient* in terms of the frequency with which such behaviors are used by an individual, and in terms of the number of ineffective actions that fail to enlist the needed assistance. If an individual is unable to enlist the assistance of the environment, he may develop a repertoire of negative behaviors to attain the desired outcome. He may request help even when it is not necessary, as a means of establishing a person-oriented dependency relationship. A patient, for example, may request the assistance of the nurse in order to establish contact, not because the patient is unable to care for himself. A child who is able to dress himself at school may refuse to do so at home in order to maintain contacts with mother. An extreme malfunction would be represented by the development of antisocial actions that could result in imprisonment or other legal punishments. A less extreme form of expressing a need for assistance to an unresponsive environment may be seen in the attempts of a child to attract his mother's attention; if he fails to do so with positive behaviors, he may succeed by hitting his sister and being punished. The desired outcome of attention is obtained through both behaviors, but with the negative behaviors the required assistance is obtained only through actions which also elicit anger and punishment.

Some individuals respond to all situations as demands that require the assistance of others. These people fail to discriminate their own abilities and those aspects of the situation which they can handle through self-reliance. Rather they have a perseveratory set to view all situations as being beyond their abilities and consequently become overly dependent on others in their environment for assistance. At the other extreme is the person who views all situations as a test of character and denies any required assistance from others. These individuals, if confronted with a situation that demands the use of dependency behaviors, such as illness and hospitalization, may find it virtually impossible to accept the expectations of the environ-ment and may become controlling and demanding.

The ability to discriminate those persons who are capable of providing assistance may also be undeveloped within an individual. The process of directing all dependency actions toward the same individual regardless of whether that individual is capable of assist-ing in all areas reflects a malfunction of the dependency subsystem in which the individual fails to develop specific help-seeking behav-iors and preparatory set. Rather these individuals attain assistance by using the discriminatory abilities of others and requiring that they enlist the needed assistance. The woman who is unable to de-cide where to turn for help may demand that her husband make the decision. In such situations the dependency subsystem of the wife is functioning only in so far as she can enlist the assistance of her husband, but he must provide the decisions and discriminations upon which specific forms of assistance are obtained.

If the environment is seemingly unresponsive to pleas for assis-tance, an individual may be left with no alternative but to turn inward and withdraw. This withdrawal may be in the form of a psychosomatic illness, or it may be a psychological disorder. Essen-tially these are maladaptive cries for help by an individual who no longer feels that there exists any means of eliciting assistance from objects or persons in the environment. The ultimate withdrawal is a suicide attempt; this represents not only a total rejection of the environment which has failed to provide assistance but also the ceas-ing of all attempts to allow the environment to assist in the future.

AGGRESSIVE-PROTECTIVE SUBSYSTEM OF BEHAVIOR

Structural Components

Goal: to protect oneself, others, or property from real or imag-ined threatening objects, persons, or ideas in the environment in order to ensure survival of the individual as a whole. Specific goals may include: directly attacking the perceived source of danger; in-directly neutralizing the perceived source of danger; withdrawal.

Set: an awareness of the presence of objects or individuals in the immediate environment that can produce injury, pain, and even death. Encroachment on one's psychological space may produce protective behaviors. The perseveratory set involves the predisposition to use the same protective patterns irrespective of the specific threat.

Choice: the range of alternatives is limited by the specific situation as perceived by the individual, by the form of the threat, and by the individual's evaluation of the degree of danger to survival. The general choices are three: (1) fight; (2) flight; (3) immobilization.

Acts: any directly observable behavior that protects the individual from injury or potential death. Specific acts include: physical fight or flight; verbal abuse; damage to property; sarcasm; avoidance; use of seat belts; and other actions that reduce the potential sources of danger that exist within the environment.

Sustenal imperatives: conditions that serve to protect, stimulate, and nurture the development of behaviors related to the aggressive-protective subsystem. Included are the learned identifications of harmful stimuli; pain perception; legal structure; ability to predict cause and effect; labeling of poisonous substances; immunizations; prior success in protecting oneself; and establishment of safe, protected environments.

Functions of Aggressive-Protective Behaviors

The aggressive-protective subsystem is comprised of those behaviors that defend or protect the individual from sources of real, potential, or imagined threat in the environment. These threats may originate outside of the individual in the physical and sociocultural environments, or they may originate within the individual in the physiological and psychological processes.

Aggressive functions related to external threat are directly linked to the perceived presence of the threatening stimulus, whereas an internal threat, such as physical pain or a state of frustration resulting from blocked movement toward a goal, may produce a disruption of previously ongoing patterns of behavior. For example, perception of an external noxious stimulus will elicit behaviors that are designed to remove the danger from the environment, if at all possible. If removal is not feasible, then actions might be taken to neutralize or avoid the potential danger, i.e., fight or flight. If a person becomes aware of a rattlesnake in the grass he may protect himself from danger by killing the snake or by withdrawing from the immediate situation. Such behaviors are specific to the threat in the environment and are available for observation by others. The protective behaviors used in a situation reflect the individual's evaluation of the situation, as well as the probability of personal risk. If the probability is low that personal injury will result from active intervention, such as when the source of the threat is viewed as being weaker in strength than the threatened individual, it is likely that an active, direct attack will be observed. If, however, the person views himself as the weaker of the two, withdrawal or passive attempts to reduce the potential harm will take place. The decision to attack or avoid is based to a large degree upon the individual's perseveratory set from past experiences in similar situations. For this reason situations that may appear to be relatively innocuous to the observer may be responded to with much more intensity of action than would be expected based upon the objective components of the situation.

An internal state of frustration results in diffuse, apparently unrelated behaviors that have little connection with the immediately preceding behavioral actions, and no apparent relationship to environmental conditions. An individual experiences frustration when movement toward a desired goal is suddenly blocked by an unexpected or unknown agent. The person may attribute the feeling of frustration to events outside himself, even though the locus of the blocking agent may be within the individual. Statements such as "It's his fault that I didn't get the raise" or "I'd be able to do that if I had all of the advantages he had," reflect a protective device of the individual that reduces the threat of failure.

These behaviors protect the individual *from* attack. The term *aggression* is often used in the opposite sense of invasion or attack. Such behavioral actions would be described as an attempt to master other elements in the environment, which is a function of the achievement subsystem of behavior. The differentiation between these two goals is relatively simple on the biological level. A bacterium or virus that invades the organism from the environment and produces a toxic state is clearly a threat to the survival of the organism; the foreign organism, on a cellular level, is attempting to master or control the environment. The production of antibodies and other immune responses constitutes a biological defense to a real or potential threat. Discrimination between attacker and defender is more difficult when one examines socioculturally based phenomena such as wars, riots, maintenance of ghettos, and segregation. These environmental conditions involve elements of attack and defense, and the distinction usually depends on the position of the observer. During World War II, for example, the internment of Jews in concentration camps was described by German officers as an attempt to protect these individuals from potential danger and harm in the form of attack; in contrast, the internment has been collectively designated by western countries as an example of a most vicious attack upon a cultural group in the history of mankind.

Implicit in the function of the subsystem is a concern with maintaining the status quo. A threat often implies a form of external pressure for change, or an impediment to an ongoing process. Aggressive-protective behaviors are designed to counteract the direction of change and are often classified as active or passive resistance. These behaviors are attempts to regain control of the situation. Control may be maintained by avoiding all situations that contain unknown elements as a means of protecting the individual. Refusal to try new foods, or to move away from known settings are actions that prevent possible harm from the unknown, while maintaining the status quo.

A major function of the protective subsystem involves the experience of pain and learned actions that are successful in avoiding situations that produce pain. A child learns to identify potentially harmful objects through the experience of pain. The immediate, instinctive response to pain is to avoid and withdraw from the source. Therefore, protective behaviors involve a large element of physical or psychological withdrawal. A major problem faced by an individual who is to undergo a planned surgical procedure is the probability that he will experience some pain in relationship to the surgery. If the anticipated intensity of pain is high the natural responsive tendency of avoidance will be active, and the individual may become extremely ambivalent as the surgery date approaches, as a consequence of the conflict between protection from anticipated pain and the necessity to expose oneself to the pain in order to effect recovery. Such behaviors might appear to be erratic, childish, or immature because of the largely undeveloped and instinctual level from which these protective responses emerge. The specific actions may occur before there has been a rational evaluation of the potential threat. This ability to act before full conscious recognition has occurred prevents the possible immobilization due to fright which would place a person in a precarious position of defenselessness while deciding the best way to protect himself.

Malfunction of the aggressive-protective subsystem involves the use of behaviors that are inappropriate, either to the specific threat or to the reality of the potential harm. These inappropriate responses often reflect a failure to realistically appraise the source of the threat. Paranoia is a symptom of mental illness in which the external environment is perceived to contain potential threat to the individual. This perception initiates a complex pattern of protective behaviors which may have little foundation in reality. Even "normal" individuals have certain areas of misperception, in which they unintentionally distort the meaning of a message. These "pockets of paranoia" are activated by situations in which the individual is uncomfortable or insecure to begin with. A person who has been invited to a dinner party and who questions the reason for the invitation may misperceive the location of the dinner. Upon arrival at the wrong restaurant he may protect himself from the disappointment by "rationalizing" the mistake, or concluding that the invitation was not really sincere and he was purposely misled. Such misperceptions and rationalizations are protective mechanisms that are malfunctional in that they prevent the individual from realizing the actual circumstances of the situation.

Phobic behaviors are protective mechanisms that result in the avoidance of feared objects or dangerous situations. Many individuals have certain areas of phobic anxiety which may not interfere with normal activities as long as the phobic object or situation is avoided. For some people, however, the anxiety related to the feared object or situation becomes incapacitating, severely restricting behavior and environmental interaction. In such cases, the phobia no longer serves a protective function but becomes instead a highly malfunctional behavioral pattern. The individual may become totally immobilized for fear of encountering the threatening object.

Failure to recognize the presence of potential threat in the environment can result in the eventual extinction of the individual. Major regulatory mechanisms of the aggressive-protective subsystem are the coping responses of denial or avoidance, vigilance, and nonspecific defenses. Individuals who deny the presence of potential threat are less able to respond in an effective manner to the actual threat when it occurs in the environment. This is a consequence of the reduced input from the psychological regulatory system which has essentially blocked perception of the source of danger. In contrast, persons who are highly vigilant in the monitoring of the environment for potential danger characteristically demonstrate intense responses to threats. Extreme denial is a state of malfunction in which a person may be totally unprepared to respond to the emergency created by the threat. Survival of the physiological system is a prerequisite to survival of the individual. Failure to ensure continued survival through adaptive aggressive-protective behaviors is perhaps the most serious malfunction possible.

ACHIEVEMENT SUBSYSTEM OF BEHAVIOR

Structural Components

Goal: to master oneself or others in the environment in order to produce a desired effect upon the environment. Specific goals include: mastery of persons, objects, and ideas in the environment; mastery of self in terms of competency and discipline; establishment of long-range goals for personal accomplishment; and the pleasure derived from the creative process.

Set: an awareness of the possibility of attaining control over oneself and one's environment. The perseveratory set involves a consistent, habitual tendency to select certain behaviors to attain mastery, a characteristic pattern of assertiveness, or restriction of achievement goals within a highly limited area of behavior.

Choice: the range of alternatives available to meet the goal of the achievement subsystem is extensive and includes: mastery of involuntary physiological processes such as defecation and urination; speech; occupational achievement; inventiveness; creativity; recognition; athletic competition; monetary rewards; self-discipline; and a development of a knowledge of one's self.

Acts: any directly observable behavior that serves the purpose of mastery of self or the environment. Specific acts include: voluntary defecation and urination; producing specific end products of achievement goals such as books, speeches, musical compositions, and paintings; attaining skills and technical behaviors specific to occupational groups; expressing personal preferences in order to attain desired outcomes; and other behaviors that require accomplishment and mastery of specific actions which have been formed into complex patterns of behavior.

Sustenal imperatives: conditions that serve to protect, stimulate, and nurture behaviors related to achievement. Included are the experience of success in the past; encouragement from peers, parents, and other significant persons in one's life; intelligence; challenging experiences; enriched stimulus environments; educational methods; rewards and punishments; grades; promotions; financial remuneration; prizes; increased autonomy and responsibility; and luck.

Functions of Achievement Behaviors

The function of the achievement subsystem is to help the individual attain mastery over himself and the environment in order to accomplish a task successfully. Achievement involves some sort of scale of excellence against which the individual can measure himself in relation to others This scale of excellence is socially and individually determined and provides the person with a frame of reference for evaluating success and failure. The context of the situation and the age of the individual influence the function of the subsystem to a significant degree. For the two year old the successful completion of toilet training involves mastery over his sphincters, an achievement of a particular task which is required for advancement into childhood and confrontation with more complex areas to master.

Within the American society much of an individual's status is derived from success within the achievement subsystem of behavior. For example, a family's socioeconomic class may be defined on the basis of two major factors: (1) the occupational position of the husband, i.e., professional versus blue-collar laborer; and (2) the educational level of the father. These two measures of status are closely related and cannot be said to constitute independent measures. Moreover they reflect the achievement behaviors of the father of the family on the assumption that most women tend to remain in the family household and attain their achievement goals through mastery of childrearing. This assumption is highly questionable in view of the increased numbers of married women who are employed outside of the home because of the pleasure obtained through the experience of achievement.

The effect of achievement is an expansion of environmental limits and a greater degree of personal autonomy. Achievement behaviors serve to differentiate an individual from others and from his environment, and provide him with the ability to control and modify the environment to his own demands. A person may define achievement in any number of ways. It may be to save the most money, corner the market of better mousetraps, or gain the Nobel Peace Prize.

Besides extending the control of the individual over his environment, achievement behaviors also extend the control of society over the environment through technological advances made by various members of society. The development of agriculture was a mastery over the environment that provided a stable food supply for the population. The textile and oil industries are attempts by man to manipulate and control the environment for his own benefit. All societies, whether industrial or agrarian, manipulate the environment for survival. This appears to be both the strength and weakness of human beings in relation to other living organisms. That man has been able to modify his environment to meet his needs has enabled him to live and adapt to widely disparate environments. Now technology is opening the possibility of survival in hostile, foreign environments such as the moon and the floor of the ocean. And yet, the mastery is in many ways an illusion, because the environment is becoming less yielding to the manipulation of man.

Most individuals have some idea, however vague or well-defined, of what will provide meaning to their lives. It may be happiness, contentment, power, fame, money, or security. These are long-term goals toward which an individual strives; they may or may not be achieved within a lifetime. In addition, people usually have specific, more immediate goals for the day or week, or even for the next few years. A freshman in college has the goal of learning knowledge that will enable him to achieve within his chosen occupation. A stockbroker may have the goal of mastering the stock market to gain his own seat on the exchange; a politician the goal of becoming President. Each of these goals requires mastery of one's own abilities as well as mastery over others. They require a commitment by the individual, a desire to accomplish that can sustain effort over an extended time. They also require that an individual make a realistic appraisal of his strengths and weaknesses, so that he can compensate for possible areas of difficulty.

An important aspect of mastery in relation to the environment is the feeling that one's personal preferences are able to influence external decisions and events in a way that is perceived by the individual to be favorable. This ability to express preferences is what is meant by the term *assertive behavior*.[12] The nonassertive individual is one who defers decision-making authority to others and permits the situation to master and determine his behavioral responses by failing to assert his personal preferences. For example, some persons never express a personal desire when asked what activities they might like to engage in for an evening. The person may make a vague response such as "anything would be nice" or "whatever you'd like to do," while inwardly feeling very frustrated that they never do anything that is fun. The reason for failure to be assertive may be a consequence of anxiety in a specific situation or with a specific individual, feelings of inferiority, a desire to avoid confrontations with others, or an inability to refuse a request no matter how unreasonable it might be. In contrast, the individual who is assertive is able to express with comfort and confidence his personal feelings and preferences, is able to appreciate the expression of personal preferences by others, and to refuse or accede to requests on the basis of personal evaluation of reasonableness. Such an individual is more able to respond to the demands of a situation in a way that maximizes the attainment of desired outcomes than the nonassertive individual whose responses are largely determined by the expressed preferences of others. Assertion training groups are designed to assist persons in becoming more comfortable with expression of personal preferences. Their popularity attests to the difficulty many persons experience in this area.

Malfunction of the achievement subsystem is reflected in the apparent inability to establish a life goal, or to attain any meaning or purpose to one's existence. Some individuals pass through life with a repetitive series of failures. No matter what assistance is provided by the environment these individuals are unable to develop sufficient motivation to sustain their drive toward accomplishment of a goal. They can be said to succeed only by failing.

A person may experience repeated failures because of an unrealistic appraisal of strengths and weaknesses. Individuals have certain talents that can be described as innate. Artur Rubenstein demonstrated a talent for music at the age of three. If he had become a garage mechanic he might have been an excellent master of the trade, but he would have not developed his innate talent. On the other hand, a person who cannot carry a tune may aspire to be an opera singer; the realistic chance for achievement within that chosen field is minimal. In such a case, society must help such individuals discover and develop their particular strengths for the benefit of themselves and society. Malfunctions of this type are fairly frequent in an achievement-oriented society such as the United States where artistic-creative and intellectual achievements are valued more than mechanical-manual forms. Persons who have mechanical-manual talents often are discouraged from developing these abilities in favor of the more valued areas of creative and academic success. Unfortunately, these individuals tend to experience a greater sense of failure and dissatisfaction with their achievements because they are prevented from developing mastery within their area of interest.

A serious malfunction of the achievement subsystem of behavior is an overemphasis on such behaviors to the exclusion of other goals. Formal achievement and mastery are important, but when they become the sole concern of the individual the delicate balance among the eight subsystems is disrupted. The stress associated with executive positions and the danger of developing illness such as myocardial infarctions, duodenal ulcers, and cerebral vascular accidents have focused concern on the relationship between achievement and illness. The high achiever comes to see all situations as a challenge for mastery and consequently fails to develop alternative adaptive approaches to the environment. The use of achievement behaviors is malfunctional in some situations, and it is important that an individual learn to distinguish those situations that do not require mastery or control.

RESTORATIVE SUBSYSTEM OF BEHAVIOR

Structural Components

Goal: to maintain energy balance throughout the behavioral system by transforming and redistributing energy according to the demands of the various subsystems. Specific goals include: restoration of a state of homeostasis; relief of fatigue, both mental and physical; renewal of energy for future activity; and recovery from illness.

Set: an awareness of a state of energy imbalance or impending state of imbalance that can be corrected or restored through restorative behaviors. The perseveratory set includes the predispostion to restore balance in a consistent, habitual manner.

Choice: the range of alternatives available to meet a restorative goal is extensive and includes the basic choices of: reduction in energy expenditure for a period of time; change in activity pattern to counterbalance the original source of imbalance; conscious disregard of symptoms of imbalance and continued expenditure of energy; modification of external situational factors related to imbalance; and provision of additional energy to the individual for supplemental distribution.

Acts: any directly observable behavior that serves to restore an individual to a state of relative balance in which no one subsystem dominates to the detriment of others. Specific acts include: rest; relaxation; vacation; sleep; painting; recreational activities; physical exertion; gardening; watching television; reading; daydreaming;

and other actions such as varying patterns of activity between sedentary and active processes, alternating required activities with desired activities, and changing environmental conditions.

Sustenal imperatives: conditions that serve to protect, stimulate, and nurture behaviors related to restoration. Included are: prior experience with effective restorative acts and the associated reduction in fatigue; economic resources; ability to recognize impending fatigue and symptoms of illness; medical and nursing personnel; adequate medical institutions in the environment; recreational facilities; ability to relax; entertainment resources; and the availability of time.

Function of Restorative Behaviors

The general function of the restorative subsystem of behavior is to restore a state of balance within the individual and between the individual and his environment. Through these behaviors an individual prevents the development of illness states that are a consequence of fatigue and overexertion. Unfortunately many people cannot relax enough to effectively replenish their energy stores. Perhaps it is a reflection of the puritan ethic of "idle hands make idle mischief" that creates such a difficulty for so many people.

Most people sleep about eight hours a night. This activity of rest is characterized by the restoration of energy balance and of energy stores through anabolic processes. Metabolism of energy is decreased to a minimum, and the body replenishes supplies for the anticipated expenditures of the following waking period. Any malfunction of the physiological factors of anabolism, such as glucose and lipid conversion and storage, will be reflected by a decreased tolerance for activity and in an ineffective relief of fatigue states. It is important to remember that sleep is an active state of inactivity. Any interference or modification of an individual's sleep pattern may affect the waking pattern of activity. The classic study by Dement of the effects of experimental deprivation of dream states had concluded that there is a restorative function to dreaming and that the behavioral consequences of deprivation included irritability, delayed reaction time to stimuli, decreased memory and ability to recall recent events, and generally disorganized patterns of behavior.[13] Johnson reports that subsequent studies have failed to confirm these findings.[14] The preponderance of evidence suggests that sleep loss of any stage may have the consequences reported by Dement and that it is the amount of total sleep time that is the critical factor, rather than the type of sleep.

Restorative behaviors also constitute an important activity during the waking hours. It is difficult for most individuals to pay attention to or to engage in one activity for an extended period of time. Prolonged periods of intense concentration may be followed by a period of physical activity. This change in form of activity allows expression of the accumulated stores of energy. Many people find painting, gardening, physical exercise, walking, listening to music, and other activities to be helpful for the restoration of mental energy stores. Singer reports that daydreaming, e.g., inner fantasies and thoughts, helps maintain an interesting environment when a person is confronted by an important but monotonous task.[15] The specific activities used by individuals for relaxation and restoration are reflective of their life styles and preferences, but can be characterized as those which provide a renewed sense of wellness and energy.

Prolonged failure to engage in restorative behaviors creates the potential for development of illness, and a period of *enforced* rest. An alarming trend is the increasing incidence of myocardial infarctions among males in the younger age groups, i.e., 30–40. A characteristic pattern of behavior among these individuals is that they are highly achievement oriented, spending long hours in intense concentration with little, if any, time devoted to diversionary forms of activity such as relaxation or hobbies. Even on the golf course or in other physical activities their behaviors tend to be achievement oriented and competitive, not restorative. They are constantly on the go until forced to rest by the development of cardiac pathology. Many corporations have become concerned with the potential loss of these young executives and have begun to develop formal mechanisms to encourage restorative behaviors in this group. Some institutions have provided gymnastic equipment for employees and encourage the attention to exercise for well being. The development of illness, in general, indicates that the restorative behaviors must become dominant for a time in order to restore a state of health. Although not all disease processes are a direct consequence of malfunction of the restorative subsystem of behavior, the failure to restore balance can contribute significantly to the onset of disease and delayed recovery.

SUMMARY. This chapter has presented the basic structure and function of each of the eight subsystems of behavior. These subsystems interact with one another to produce complex behavioral patterns by which the individual relates to his environment. Behaviors commonly associated with one subsystem may be used by an individual to achieve a goal associated with a different subsystem. For example, a person may use eliminative behaviors for the expression of affective states to master his environment, rather than to simply communicate to the environment the existence of an internal feeling. It is part of the uniqueness of each individual that he can utilize any number of personally valued behaviors to attain a desired outcome. Those which have been identified in this chapter as associated with a particular subsystem of behavior reflect the general function of the behavior for most individuals. It may not, however, be valid for all. It is for that reason that a comprehensive assessment of the various components of the behavioral system should be conducted for each individual.

The following chapters discuss the regulatory functions of the external environment as well as the physiological, psychological, and sociocultural factors that regulate activity of the behavioral system. Chapter 5 discusses the environmental factors that influence behavior and the process of perception by which an individual transforms the external environmental stimuli into a subjectively personal event. Chapter 6 outlines the general regulatory factors that apply to all eight subsystems of behavior.

STUDY QUESTIONS

1. Observe the behavioral pattern associated with ingestion of food for two individuals, one who is ill and hospitalized, and one who is healthy. Indicate those component actions of the behavior that are similar and those that are different. How does the situation affect the behavior?

2. Repeat the above assignment for each of the remaining seven subsystems of behavior.

3. Select a patient whose disease process involves a malfunction of one of the subsystems of behavior. Identify the elements of the behavioral pattern that are adaptive and those that are maladaptive.

4. Observe one individual in a variety of formal and informal groups to which he or she belongs. What are the criteria for affiliation within these groups? How does the individual's behavior differ qualitatively when the interaction involves a friend versus an acquaintance.

5. Observe a person who is talking and identify the verbal and non-

verbal behaviors associated with the communication process. Are the two sources of information conveying the same message?

6. Observe a patient within the hospital setting and describe his or her dependency behaviors in terms of seeking: assistance; reassurance; physical contact; recognition.

7. What sustenal imperatives in the hospital environment serve to nurture, protect, and stimulate restorative behaviors?

five | *The External Environment as a Regulator of Behavior*

An individual is constantly exposed to an extensive variety of stimuli that exists in the external environment. It is impossible for a response to be made to all potential stimuli, because it would exceed the ability of the human organism to process all the information rapidly enough to produce a coordinated response. As a consequence there must be some mechanism to regulate the processing of incoming stimuli so that those messages that have personal meaning to an individual or that are necessary for continued survival of the organism are recognized, while other stimuli, which are meaningless, are ignored. The process of perception, whereby an incoming stimulus is registered and transformed into a subjectively meaningful message, regulates the amount of contact between the external environment and an individual. This chapter identifies the major stimulus forms that exist in the external environment and the major determinants of the perceptual transformation of these stimuli into subjective, personal messages. These transformed stimuli constitute the basic input to the behavioral system, which initiates the process of selecting an appropriate response pattern.

Input that originates from the external environment refers to the information contained in the form of persons, objects, and situations that are external to the boundary of an individual. A second, equally important source of input originates from within an individual in the form of emotions, motivations, thoughts, fantasies, attitudes, beliefs, and reflexive responses. An individual must apprehend and evaluate these complex stimuli for their meaning and intent. This evaluation process involves the activity of the biological, psychological, social, and cultural regulators and their contribution to the transformation and differentiation of the stimulus into an internal, personal event. The external environment and the bio-psycho-socio-cultural aspects of the individual are considered to be *regulators* of behavioral system activity because they directly determine the specific information upon which the response is based. The relatively stable characteristics of an individual, such as personality traits, social status, and other determinants, are regulatory factors that are present in all situations. These general regulators are discussed in Chapter 6. Characteristics of the situation are regulators which influence whether a response is made to a particular stimulus and includes time; current level of adaptation; expectancy; and attentional determinants. A third group includes regulators specific to the behavior which they tend to elicit. These are phenomena which have a high probability of eliciting a specific behavioral pattern over an extended period of time and over a wide variety of situations; for examples, the association of hunger pains and gastric contractions to the ingestion of food; and the relationship of ability, task difficulty, luck, or effort to the use of achievement behaviors.

The end product of perception that constitutes subjective reality for the individual may or may not correspond to objective reality. The entire process, from moment of occurrence of the event in the external environment to completion of the associated response requires very little time or energy for the normal, healthy adult. Any pathological factor, such as brain damage, depression, an unfamiliar situation, anxiety, sensory loss, a new cultural setting, or extreme fatigue, will affect eventual integration and response by the behavioral system. Responses tend to become fragmented and disjointed, and require greater amounts of energy than may normally be required for the simplest action. The extreme point is exemplified by the pathology of schizophrenia in which responses of the individual are based upon internally generated stimuli and an association to the external environment is absent. A similar situation can be produced in healthy individuals who are placed in a severely restricted sensory environment.

THE EXTERNAL ENVIRONMENT

The external environment includes all those persons, objects, and phenomena that can potentially permeate the boundary of an individual. The assumption of potential permeability is of vital importance, for if a phenomenon cannot enter the system, thereby making itself known, it does not exist for an individual. People who are colorblind to red and green, e.g., deuteranopes, are unable to perceive these colors even when a light restricted to that portion of the color spectrum is presented to them. For these people, the colors of red and green do not exist.

It is possible to describe the multiple dimensions of the external environment on many abstract levels, both qualitative and quantitative in nature. Specification of the external stimulus requires identification of the physical event, or change in physical energy, that forms an organized or meaningful pattern and that elicits a response from the individual. There are many energy changes occurring in the external environment at any given time that do not constitute stimuli because they fail to evoke a response. These are random, chance stimuli that form a background matrix upon which the meaningful stimuli are superimposed. The absence of a response, when one is expected, may result in a tentative conclusion that the individual has failed to apprehend the critical event in the environment. The expected response need not occur in close contingency with the eliciting event, since the process of retention of input, such as memorization and recall, allows for an extensive time lag between the occurrence of original input and subsequent response. However, there is almost always some form of behavioral response available to the observer which indicates that the input has been received, regardless of whether the specific pattern of response takes place. Such nonspecific indicators as eye contact, nodding of head, and associated nonverbal actions that communicate simple comprehension may be used to evaluate the success or failure of communication.

Once the critical stimulus has been identified, it is possible to describe it as a complex phenomenon composed of x dimensions. An object can be described in terms of length, weight, color, temperature; dimensions of the physical environment include intensity, variability, temperature, weight, size. People can be dimensionalized in terms of sex, age, height, weight, race, political affiliation, religion, occupation, and so forth. The difficulty of the task is to specify all those dimensions of the environment or of the critical stimulus that affect a given individual and directly activate corresponding behaviors within the individual. The following discussion briefly describes the general stimulus dimensions of physical settings, persons, objects, and psycho-socio-cultural attributes of the external environment.

The Physical Environment

The physical environment contains a wide variety of potential stimuli, chief of which are the physicochemical energy factors associated with sensory phenomena, such as light and sound waves, and the meteorological stimuli. Many dimensions are associated with the meteorological and climatic stimuli, all of which have assumed greater importance for man as the interdependence of life with the environment has been increasingly emphasized. Included in this broad class are the following dimensions: temperature variations; wind velocity; atmospheric pressure; absolute and relative humidity; solar radiation; air pollutants; ozone, oxygen, carbon monoxide, and carbon dioxide levels; electrostatic and electromagnetic fields; sunspot activity. These stimuli exhibit variability on two basic dimensions of time: (1) 24-hour variation related to the day-night periodicity; and (2) lengthier, more gradual variations related to the yearly seasonal changes.

Meteorological stimuli have important physiological, sociocultural, and psychological attributes. The physiological effects are mediated primarily by the hypothalamus, in particular the thermoregulatory centers of the anterior and posterior hypothalamus, and the autonomic nervous system. The hypothalamic factors are of primary importance in compensating for seasonal variations, whereas the autonomic factors maintain a homeostatic equilibrium in the constantly changing environment in which an individual is physically located at a given time.

Seasonal variations are multiple and complex, directly or indirectly involving every physiological system. The following effects have been found to be related to seasonal changes in the winter months, as compared to the summer: In the winter, calcium, magnesium, and phosphate levels in blood plasma are lowered; thyroid and adrenalcorticoid activities are elevated; hemoglobin levels are increased; and gastric acid secretion is high.[1] These changes are gradual in nature and reflect the *thermostatic* properties of the hypothalamus in the conservation of heat and energy during the winter months when the external environment is cold. During the summer, when the general weather pattern is one of environmental heat, the thermostatic levels of the hypothalamus favor heat loss, since less energy is required to maintain a required level of internal heat. It has been suggested that this thermostatic property accounts for the subjective experience of warmth or coldness that is independent of the absolute environmental temperature but is related to the seasonal average. A temperature of 50°F during the winter, following an extended period of freezing temperatures, will be experienced subjectively as a warm day, while the same temperature following a period of 90°F temperatures will be experienced as cold. In the first set of conditions the body is prepared to conserve heat and the counter trend of the weather is experienced as a psychophysiological contradiction.

Superimposed on the seasonal foundation are the daily variations of thermal stimuli. The more extreme the change in weather, the greater the response. For example, the response to an influx of a cold front includes: diuresis under conditions of constant fluid intake; increased thyrotrophin production; elevated leukocyte and thrombocyte levels; elevated hemoglobin; lowered erythrocyte sedimentation rate; and increased fibrinolysis. The opposite changes occur during a sudden heat spell. These physiological adaptations to the environment are mediated in such a manner that under most environmental situations a relative state of constancy is maintained by the autonomic nervous system. Any extreme fluctuation in environmental conditions or the sudden change of environment by an individual will result in a temporary disruption of the internal environment, thus requiring the use of increased levels of energy to restore physiological adaptation.

The sociocultural attributes of weather are reflected largely in the life style of man in relation to his physical environment. Clothing during warm weather is designed to facilitate the radiation of heat from the skin to the surroundings, whereas cold weather requires heavier clothing as a supplementary insulation layer to reduce the radiation of heat to the environment and, instead, to retain the heat next to the skin. Thus, during the summer clothing tends to be loose, scanty and in material of light colors that facilitates reflection of light and heat. Food intake tends to decrease and the types of food preferred reflect reduced requirements for heat production by the body. In the United States, the warm months encourage less formal life styles, with many of the activities being related to the outdoors, e.g., camping, tennis, swimming, fishing, etc. However, sociocultural factors may gain a large degree of independence from actual weather conditions. For example, in the past women's winter clothing was primarily of dark color, but in recent decades the lighter, more saturated hues have been available year round. As more and more people live and work in temperature-controlled environments in which the seasonal variations can be controlled by a thermostat setting, it can be predicted that the sociocultural factors associated with weather variations will become less marked.

The psychological attributes of weather and climate are primarily based on an individual's personal preference. Many people experience a feeling of well-being, increased energy, and efficiency during temperate conditions. Prolonged periods of fog, rain, or snow are often associated with a feeling of depression and a reduced ability to think clearly. Solar radiation and ionization factors have been implicated in the experience of irritability and tension during periods of unstable atmospheric pressure, high winds, and low heat energy. In Hawaii such periods of intense instability have been culturally assimilated by the native Hawaiians as a "devil wind," during which death and destruction may occur; this represents a symbolic interpretation of weather conditions over which man has little control.

An extended heat wave during which the temperature was above 90°F for 14 consecutive days in Los Angeles provides an excellent example of the relationship of weather to physiological, psychological, and sociocultural regulators of behaviors. Experts from a variety of social agencies were interviewed.[2] Dr. Charles Wahl, a psychiatrist, noted that "a sustained period of heat shortens our tempers, makes us irritable and depressed. There is a difficulty in sleeping so our emotional resources tend to be used up." Depletion of salt reserves, decreased appetite, and slowed metabolism contribute to the feeling of lethargy and constitute a very real state of decreased energy. Police reported an increase in domestic quarrels and crimes, particularly an increase in crimes involving other persons rather than property. There was also an increase in arrests for vagrancy since the chronic drunks would sleep in an alley rather than seek shelter during the heat wave. Demands for ice, beer, and ice cream resulted in record-setting sales. It is usually only during extreme weather conditions that most people become consciously aware of the effects of environment on their own functioning. Nevertheless, these same stimuli are constantly providing important sources of input to the physiological, psychological, and sociocultural systems, and thereby producing specific patterns of responses within the individual.

Whereas weather and climatic conditions form a distal matrix which is applicable to all individuals within a given geographical setting, the immediate physical environment also provides multiple sources of stimuli that are more specific to the setting. The design of a room together with the persons and objects located in the room combine to form a gestalt that may be termed pleasant or unpleasant. The colors of the room, for example, may produce a feeling of quietness, of coolness, or of nature by the use of forest greens, light blues, or browns, whereas warmer colors reflect a feeling of heat and intensity. The significance of environmental colors was acknowledged with the introduction of green into the operating room setting, in place of aseptic white. The wearing of colored uniforms by nurses and the introduction of colors and decorative elements into hospital rooms further acknowledge the impact of the environment upon the response of an individual to hospitalization. Air movement and room temperature also influence associations to the ambience of a setting. Persons situated in a room in which the ambient temperature is high with little or no air circulation have been found to express negative associations to persons and objects located in that environment.[3]

Sensory stimuli located in the environment are physical and chemical processes that can effect neural activity in specific physiological receptors. These stimuli constitute a major portion of input from the physical environment and are the primary source of an individual's knowledge about his physical and social environment. The sensory stimuli serve as basic building units that form an infinite variety of complex stimuli, such as combinations of sound and light, texture and taste. These complex patterns of stimuli provide the matrix of potentially meaningful, specific information for the individual. Physical energies exist in the environment at all times, but are known only when they are able to penetrate the boundary of a system that can perceive their existence. The question of whether color is possessed by an object or by the person perceiving the object is an expression of the complex relationship that exists between the physical stimulus and the psychological representation of the physical event.

Study of the direct relationship between the specific properties of the physical stimulus and the perceived quality experienced by the individual is called *psychophysics*. Psychophysics also is concerned with the laws governing the quantity of available stimuli perceived by an individual. The organism is not sensitive to the entire range of physical stimuli which exist in the environment. During normal conditions some lights are of too low an intensity to be seen, some sounds are too soft or of a frequency that is not audible to the human ear, and so forth. The *absolute threshold* is that point in the scale of physical stimulus intensity where the stimulus just begins to be perceived by the individual. In terms of systems analysis, the absolute threshold is that point where the filtering mechanism of the boundary allows passage of the stimulus through the boundary of the system. This threshold is fairly stable over time for the same individual, but will vary under differing stimulus conditions, as well as in different conditions of measurement. Measurements of the absolute threshold are obtained under highly controlled conditions, during which attempts are made to eliminate all other stimuli that would interfere with the perception of the critical stimulus. Obviously this is quite different from the normal, uncontrolled setting in which stimuli are perceived on a day-to-day basis. The entire field of psychophysical measurement is too complex and extensive for a text of this intent. The following material is concerned with the basic parameters of sensory phenomena as they exist in the form of physical energies in the environment.

The physical stimulus for visual experience is light, which can be defined as radiant energy which originates from a primary source, such as the sun, a candle, or a light bulb. This energy has been shown to have two basic forms: (1) a stream of particles, corpuscles, or discrete quanta; and (2) a spectrum of waves of varying lengths. Each of these two theoretical forms has different properties, and yet each is able to produce the visual experience of light.

Light can be described in terms of two basic dimensions: *intensity*, defined as the amplitude of the stream of quanta or of the wavelength, and measured in standard units of international candles;[4] and *wavelength*, defined as the inverse of the frequency/speed ratio of the stream, measured in units of millimicrons (mμ).[5] The human eye is not sensitive to the total spectrum of wavelengths existing in the external environment. The visible spectrum, or that portion of the total spectrum to which the eye is sensitive, is fairly narrow in range, (724 mμ–397 mμ), and is bounded by the long infrared heat waves and the short ultraviolet waves. Most light within the visible spectrum is a mixture of wavelengths in which the dominant wavelength present serves to define the *hue*, or color, of the physical stimulus. Although most colors appear to be pure on a perceptual level, the physical stimulus always includes wavelengths other than the wavelength which is associated with the perceived color. For example, red represents that portion of the visible spectrum between 723–647 mμ wavelength. Wavelengths of other portions of the spectrum will be present also, in varying proportions, producing a degree of desaturation. As the proportion of wavelengths other than the dominant hue increases, the color becomes increasingly desaturated, or washed out. The maximal desaturation that can be attained is represented by white, at which point no one wavelength is dominant. The third characteristic on which colors can be discriminated relates to *brightness*, which corresponds to the intensity of the radiant energy. The greater the intensity of the light, the brighter the color— i.e., darker colors are less intense and have a smaller amplitude of the stream of quanta than do brighter colors.

Sound waves constitute the physical stimulus for the perceptual experience of hearing. The waves are longitudinal in nature, and consist of molecular motion in the direction of energy transmission which occurs in an elastic medium such as air or water. A simple sound wave, which is the basic unit of all complex sounds such as speech and music, consists of an alternating compression (condensation) and expansion (rarefaction) of air particles. To penetrate through the boundary of the individual, these waves must impinge upon the eardrum and establish a related vibratory pattern.

The two main characteristics of sound are amplitude and frequency. *Amplitude*, or intensity, is defined as the total energy in a wave measured by a logarithmic scale of decibels,[6] or alternatively, measured as the peak pressure of the sine wave. *Frequency* is defined as the time period required for a complete sine wave cycle, measured in the number of cycles per second (cps). These two dimensions, frequency and amplitude, serve to describe all simple and complex sounds. Speech consists of many different frequencies occurring simultaneously, and it is only through the application of Fourier analysis that the complex frequencies can be differentiated into the simple units. The pitch of a sound will vary with the frequency of the wave. Low tones are evoked by low frequency waves, while high pitches are related to high frequencies, although the relationship is not a pure one-to-one phenomenon. *Loudness*, the perceptual dimension that ranges from a barely perceptible hum to an earsplitting siren, is related primarily to amplitude. The pitch of a tone may affect the perceived loudness, for if two tones of equal amplitude but different pitch are played they will not be perceived as being of equal loudness. The meaning of physical stimuli is established through periodicity, or purposeful repetitiveness. Nonperiodic vibrations, characteristic of the majority of sound waves in the environment, constitute the basis of "background noise."

Sound and light are both active currents, or streams, of stimuli moving from one point in space to another. Thus they convey information concerning their point of origin, which may be distantly situated in both time and space from the point of perception. The light of a star reaching earth and registered by the eye was emitted from that star decades ago. The information contained in such light represents the conditions present at the time the stimulus was emitted, and a like period of time must pass before it is possible to describe the conditions of the stimulus source on the day of the original measurement. Other forms of physical stimuli are more localized in the environment, and provide information about immediate conditions surrounding the boundary of the individual.

Chemical stimuli in the physical environment serve to initiate activity in the form of taste and olfaction. Four pure types of chemical substances have been identified as underlying the complex sensations of taste and are perceived as: (1) bitter; (2) salt; (3) sweet; (4) sour. The chemical basis of *sour* is the presence of hydrogen ions in an acid solution. *Salt* may be elicited by a number of dissimilar ionic compound substances, but it is believed that the negative ion is the critical factor. The perception of *bitter* can be produced by alkaloid substances and inorganic salts, including quinine. The most common physical stimulus for perception of *sweet* is a sugar substance, although nonsugar compounds, such as sodium chloride, may taste sweet at certain concentrations. Each of these chemical substances must come into direct contact with the boundary of the perceptual system, in this instance the tongue and other parts of the mouth and throat, in sufficient concentration for the experience of taste to be elicited.

A second form of chemical stimuli—rapidly diffusing molecules of volatile substances—initiates activity when these particles come into contact with olfactory receptor cells located in the nasal cavity. Adrian tentatively identified four classes of chemical substances that are capable of initiating olfaction.[7] These are: (1) acetones and ethereal esters; (2) aromatic hydrocarbons such as benzene; (3) paraffin hydrocarbons and heavy oils; and (4) terpenes and related substances such as cedarwood oil and eucalyptus oil. These volatile substances must be suspended in air for the experience of smell to occur. It has been reported by Krech and Crutchfield that an investigator who filled his nasal cavity with eau de Cologne was unable to smell any of the aromatic contents.[8]

The physical stimulus for cold and heat is fluctuation in the environmental temperature. Cold is related to a drop in temperature at or exceeding the rate of 0.004°C per second; warmth is signalled by a rise of at least 0.001°C per second. The significant factor related to the perception of cooling or warming is the temperature immediately preceding the change in temperature. Perception of cold, or a drop in temperature, is possible only if the preceding temperature was higher, i.e., warmer. The absolute level of the environmental temperature is a contributory factor to the experience of variable temperature.

All objects, events, and persons in the environment can be described in terms of complex sensory stimulus dimensions. Objective description of an object involves the formal and informal measurement of the critical factors that serve to define the object under study and to distinguish it from all others. It is possible to measure the objective dimensions of size, weight, mass, and temperature. The boundary of an object, when held in the hands or touched, exerts pressure upon the boundary of the physiological system—the skin—resulting in active stimulation of mechanoreceptors in an organized pattern. This pattern of pressure is then transformed, through the complex process of perception within the central nervous system, into a psychological description of the boundary shape and the degree of firmness of the object. Simultaneously, other properties, such as temperature, color, and brightness, are registered by the appropriate sensory receptors. These independent sources of information are integrated eventually into a psychological representation of the original external object. This entire perceptual and cognitive process may be completed in an instant, thereby blurring the individual bits of information that have been used to produce the inner representation. Once an object has been internalized, it is then compared to other objects that have similar and/or dissimilar dimensions. This process of comparing the presently perceived object to those perceived at earlier times enables an individual to relate to the object as a member of a broader class of objects with known function and structure.

It is the end result of the perceptual process, the labeled internal perceived object, that constitutes the immediate input to the behavioral system. Responses and actions of the individual are directed to the inner perception of the outer reality, and as such are subject to individual interpretation. No two witnesses to a crime will report the identical event. One may have focused on the clothing worn by the attacker, another on the facial emotion of the victim, and a third on the events surrounding the crime. When aspects of the situation or object or person are concrete and can be formally measured, inner perceptions can be communicated by using these measurements. The difficulties are great, however, when the perceived event contains few measurable aspects. Discussion then becomes dependent upon the inner psychological representation which is subject to individual bias, distortion, situational variables, affect, and so forth. It is important to remember that *the observed response will be congruent to, and determined by, the individual's perception of the reality, not by the reality itself.*

The Psychological, Social, and Cultural Environments

The psychological environment surrounding the individual is more difficult to objectify for it is constituted by those aspects of the setting that have acquired a specific psychological meaning for a given individual. Psychological ecology is the field of study of individuals and groups in interaction with the psychological environment in which human beings grow, mature, and behave.[9] The way a person perceives his surroundings is a major determinant of the way that individual will behave. Environments can shape adaptive potentials and can facilitate or inhibit strategies for coping with the various elements contained in the setting. Moos discusses psychological environments in terms of three primary dimensions:

1. relationships
2. personal development
3. system maintenance and system change.[10]

Relationship dimensions refer to the nature and intensity of personal relationships among the various individuals and the ways in which the environment fosters support, involvement, and interaction among the participants. The personal development dimension reflects the potential or opportunity which exists in the environment for personal growth and development of self-esteem and self-worth. The third dimension of system maintenance and change indicates the extent to which the environment is orderly and clear in its expectations, the degree of control exerted by the environment upon the various individuals, and the responsiveness of the system to proposed change. Each of these dimensions is present in all psychological settings and will function, in varying degrees, to facilitate or inhibit behavior.

There is an infinite variety of situations to which an individual may be exposed, but over time there emerges a finite group of basic situations that share a common element. A child may experience an infinite number of classrooms and teachers, all of whom are distinctly unique, but the situations share the common factor of a teacher-student relationship. Throughout one's life, objects, persons, and events become associated with positive and negative values. Those that are positively valued by the individual will be sought out actively, and will serve to direct and motivate behavior toward their attainment. The birth of a child, accumulation of great fame or wealth, good health, and self-actualization are examples of positive, sought-after situations. Those factors that are negatively valued will initiate avoidance behaviors directed to attempts to reduce the potential influence of these environmental factors upon behavior. Except under rare circumstances, people view death as a negative event. A person learns to avoid, as much as possible, all objects, situations, and people in the environment that can produce this state of death.

The sociocultural environment includes other people and groups; social institutions such as political, economic, and educational systems; social events such as war, peace marches, political campaigns, debutante balls; and shared moral, ethical, and religious beliefs. All groups have developed rules and regulations to assist an individual in the process of becoming a useful and valued member of the group. These rules relate to the areas of family and kinship, residence, religious beliefs and values, social situations, child rearing, marriage, death, occupations, and so forth, and serve as constraints on the range of possible behaviors. An individual who is in his native sociocultural environment most likely will behave according to the societal rule for the particular situation in which he finds himself. Society helps an individual to decide upon the rules applicable in a particular situation, but it also grants him the privilege of ignoring the rule if he so desires. A person at a formal dinner party may be much more comfortable eating chicken with his fingers and do so with the knowledge that a rule of etiquette is being broken on purpose. Punishments may be established if the rule that is transgressed is of particular importance to the survival of the group. The legal system establishes different penalties for illegal actions based on the presence or absence of premeditation, e.g., first degree murder versus accidental homicide. Every child grows up in the sociocultural environment knowing what social choices will be available to him in relation to occupation, marriage, recreation, and so forth. Every family and social group teaches its members those skills and behaviors believed to be essential for the individual to survive and to contribute to the continued survival of the group.

The rules, values, and beliefs of the sociocultural environment are fairly stable and resistant to change. This stability allows the individual a variety of choices with a reasonable basis for predicting the possible outcome. When rapid change does take place there is a period of intense social instability, where old values and relationships no longer serve to relate the individual to other members of the group. Such a situation prevails at this time in the United States. Behaviors that have long been morally and ethically proscribed, such as abortion and unwed mothers retaining guardianship of their children, have begun to become part of accepted social practice. The previously accepted moral standards of premarital chastity and childbirth within the sanction of a legally binding relationship of marriage have come into direct conflict with the social values of the young adult. The potential consequence of instability in the sociocultural environment is conflict and alienation between groups, and a lack of direction for the socialization of future members.

The effect of population density on social behavior is of increasing concern, particularly as the problem of world overpopulation in relation to available resources becomes critical. Calhoun allowed a group of four female and four male rats to populate in an "ideal" environment, defined as one that contained no known natural limits on population such as illness, famine, and extreme weather conditions.[11] As the colony increased in density the increased contact between animals became so intense that confusion developed. Soon the number of available social roles were fewer than the available mice. The observed consequence of the overpopulation was social withdrawal and avoidance of other mice. The females lost interest in tending their young, which resulted in fewer mature rats and consequently decreased breeding. The end result was sterility and eventual extinction of the colony. The parallels of this study to the description of life in the overcrowded ghettos of cities and underdeveloped nations are intriguing.

A person who travels and is exposed to a sociocultural environment different from his own can gain a feeling for the ways in which societies differ and how they expedite the relationship of the individual to the group. In primitive societies where the survival of the individual and the group is determined by basic physiological requirements to a greater extent than in a technological society, group-oriented behaviors that directly contribute to the well-being of the group are encouraged and sanctioned by members—farming and hunting for example. The individual in a technological society has less dependence on the group for such basic requirements as food and is more able to develop innate talents that may have personal value only. The focus of Eastern religions on a meditative life as a means of transcending the earthly existence reflects a very different orientation toward acquisition of material possessions, compared to Western religious beliefs.

An individual is a unique product of the interaction between his personal potential and the society into which he is born. It is only in rare instances that an individual can become assimilated into another cultural group, for the influence of early socialization and enculturation remains high throughout life. The experience of the Japanese community in the United States indicates that it requires three generations before total assimilation of the new cultural values occurs. The first generation, or immigrant group, tends to maintain the cultural rituals and values associated with the culture from which they emigrated. The second generation is socialized by both cultures, although the rituals and values of the family tend to be dominant over the values of the socializing agents of the society, such as schools. By the third generation, however, the enculturation is determined by the dominant cultural setting, and conflict arises between the values of the original culture, which are still adhered to by the parents, and the values of the child. Conflicts of this nature are difficult to resolve, for the cultural values of an individual are the primary determinants of his personal belief system and life philosophy and are often resistant to discussion on a rational, logical level.

In summary, the environment surrounding an individual contains a wide variety of sources of information, all of which are potential forms of input into the behavioral system. A person is not isolated from his surroundings, either in the form of physical stimuli, persons, objects, or events that take place around him. Rather there is a definite interaction between the setting and the behavior of an individual. The society and culture in which he is raised provide a structure for relating to members of the group; the psychological environment transforms the rules into a personal experience that has meaning to the individual. The observer of behavior must attend to all the factors of the environment that are potential elicitors of behavior, as well as to the response itself.

REGULATORY FACTORS RELATED TO THE PERCEPTUAL PROCESS

Somehow the information contained within the environment that surrounds an individual and that is required for continued survival must be made available to the behavioral system. The dynamic process by which this occurs is called *perception*, an internal process that is primarily controlled by the sensory receptors, but is influenced by many other factors such as attention, past experience, set, emotional affect, and others. Since all stimuli can potentially penetrate the boundary of an open system there must be some selectivity by the system, which allows some stimuli to be perceived and attended to, while excluding others that are of sufficient intensity to be perceived. This selective process must occur prior to the conscious awareness of the perception, and can be assumed to be outside the conscious control of the individual.

There are many factors that serve to determine the final, transformed image of perception. For this discussion, these factors have been divided into five classes. It should be remembered that the process of perception, from onset to completion, requires only seconds or milliseconds, and that all of these factors operate in such a way that they are normally indistinguishable from one another. The five classes of regulatory factors are those related to:

1. transformation of the physical stimulus into a neural event;
2. the process of attending to, or seeking input from, the environment;
3. the process of exclusion of stimuli from consciousness;
4. the interpretation or cognitive meaning of perceptual contents;
5. general responses, or correlates, of perception.

Transformation of the Physical Stimulus into a Neural Event

Specialized receptors are located at, or near, the boundary of the physiological system and the environment, and serve to receive physical stimuli and transform them into an internal message. Each receptor is especially sensitive to one form of energy over all others. The particular form of energy to which a receptor is most sensitive is termed the *adequate stimulus*. Light rays within the visible spectrum constitute the adequate stimulus for vision; mechanical pressure and deformation of the paccinian corpuscle in the skin is the adequate stimulus for the perception of pressure. This is a differential sensitivity, however, for given another energy form of sufficient intensity the receptor can be triggered off into activity—e.g., pressure on the eye can, if sufficiently intense, trigger off a visual experience. When this occurs the sensation will be related to the receptor which has been stimulated rather than to the energy form of the stimulus. Johannes Müller formulated the law known as the doctrine of *specific nerve energies* which states that regardless of the physical nature of the stimulus, the subjective sensation is always the same.[12] Therefore the nature of the perception is determined by the characteristic of the receptor, not of the energy form. A simple demonstration is to exert pressure on the side of your eyeball while your eyes are closed. With sufficient pressure, the sensation of light will be experienced.

The process by which the receptor initiates the nerve impulse in response to a physical stimulus depends on the nature of the stimulus energy and the receptor. Receptors for vision, for example, are the rods and cones embedded in the retina of the eye. Light is focused on the retina and initiates a chemical process involving the decomposition of rhodopsin in the rods, and a similar substance, believed to be iodopsin, found in the cones. At a certain point in the process, dependent upon the light intensity and the amount of photochemical substance present, the bipolar cell dendrite becomes depolarized and a neural impulse is formed. Some retinal cells respond to the onset of light while others respond only when the energy source is eliminated. Sound, as a physical stimulus, initiates vibration of the eardrum which results in vibration of the ossicles located in the middle ear. The frequency of ossicle vibration is then converted into a liquid pressure wave in the cochlea, which moves with a frequency related to, but not identical with, the original physical stimulus. As the wave progresses, hair cells in the basilar membrane are bent, and if they are sufficiently deformed by the wave, this results in depolarization of the neuron and initiation of impulse formation. Once an impulse is formed it is impossible to differentiate the specific content of the stimulus message in one nerve from that in another, until decoded at the cortical level.

This process—the formation of a neural impulse in the retinal cells or in the hair cells of the basilar membrane—is similar to that of all sensory modalities. The physical energy is able to effect an initial state of depolarization in the appropriate receptor and thereby initiate the formation of a neural impulse. This impluse is transmitted over the cranial nerves or major spinal tracts, e.g., lateral and ventral spinothalamic tracts, to the thalamus at which point it forms a synapse with thalamo-cortical fibers to specic locations in the parietal cortex as well as to nonspecific projection areas. Initial perception takes place at the thalamic level, but for an integrated and coordinated response to occur it is necessary for the cortical level influences to be activated.

Factors that Facilitate
Attention to Input from
the Environment

Many factors determine whether the impulse traveling over the classical sensory pathway to the cortex will be perceived. For perception of the incoming message to take place, it is necessary that the individual be alerted to, and attend to, the incoming message. Otherwise the content may be lost to conscious awareness.

As the sensory fibers ascend, collateral fibers are given off to the reticular formation, a collection of multidendritic cells located throughout the core of the brain stem. The reticular formation, therefore, receives input from all sensory modalities simultaneously. The effect of this multiple channel input is that the specific sensory meaning is lost. These collateral fibers initiate activity in the *ascending reticular activating system* (ARAS) which alerts the cortex to the incoming sensory data over the classical pathways. The purpose is to heighten the level of activation of the cortex and prepare it for processing of the impinging stimulus. If ARAS activity is diminished due to sleep, drugs, anesthesia, or a lesion, incoming sensory messages will not be responded to and are behaviorally nonexistent for the individual, because the cortex was not alerted.

Activity within the *diffuse thalamic projection system* (DTPS), which is a collection of fibers that originate in thalamic nuclei and radiate throughout the entire cortex, serves to modulate attention over a shorter, more variable time scale. Once the ARAS has alerted the cortex that something is coming, it is still necessary that attention be focused or directed to the specific content. These two systems are anatomically connected to one another, so that the level of activity within one is related to the level of activity in the other.

Attention results in a narrowing of focus, a selective concern with certain parts or aspects of the situation and an exclusion of other aspects due to nonattention. When a person becomes engrossed in a novel, the field of perception and attention becomes narrowed to the book, or more specifically, to the page being read at that moment. All background noises are excluded from conscious awareness as a function of the properties of attention.

Certain properties of the stimulus serve to demand attention. The intensity of the stimulus, for example has a direct relationship to attention. The more intense the stimulus, the greater the number of fibers activated, and the higher the level of activity found in the two alerting and attending systems. Repetition of the stimulus, isolation from background stimuli, movement, novelty, and incongruity are other properties of stimuli that favor attention.

Related to the alerting function of the ARAS is the nonspecific *orienting response* (OR) to incoming stimuli. Sokolov suggests that orienting responses elicited by relatively innocuous stimuli are related to the individual's heightened sensitivity to environmental stimuli which leads to increased information intake and consequent learning.[13] Studies have demonstrated that if an alerting stimulus, such as a bell, or tap, precedes the presentation of a specific stimulus with a sufficient time lapse to initiate ARAS activity, the response to the specific stimulus will be of a shorter latency than if no alerting stimulus was presented. The orienting response includes pupil dilation, cardiac deceleration or acceleration, and a drop in skin conductance, the last representing a general state of arousal. Lacey and associates have demonstrated that attentive observation of environmental events is accompanied by cardiac deceleration and a drop of skin conductance levels, whereas mental work, such as mental arithmetic, results in cardiac acceleration and a drop in skin conductance.[14, 15] Tasks that require both activities result in little or no change in heart rate while other autonomic variables move in the direction of apparent sympathetic nervous system activity. The demonstration of conditions under which the change of heart rate is consistently in the direction of apparent parasympathetic activity while other variables shift toward sympathetic activation has been termed *directional fractionation* by Lacey. Repeated exposure to the stimulus that evokes an orienting response demonstrates rapid habituation—i.e., decreased intensity of ORs—if the stimulus is weak, expected, and/or not meaningful.

The quantity of stimulation present in the environment affects the sensitivity of the individual to incoming sensory information.

Studies of sensory deprivation and sensory monotony, where an individual is placed in a situation where stimulus sources are absent, or where he is exposed to an environment with little change or novelty in stimuli, have demonstrated that there is a minimum level of stimulus variability below which behaviors become disorganized. If an individual is deprived of stimuli for a sufficient length of time, internally generated sensory states, such as hallucinations, may develop. A person who has been in a darkened room for an extended period of time or a patient with an eye-patch may have periods of disorganized hallucinations. In general, the greater the absence of specific stimuli, the more heightened the sensitivity to subsequent stimulation. Removal of eye-patches and exposure to light of normal intensity may produce an experience of pain if the individual has been deprived of stimulation for an extended time. A patient in a darkened room at night will perceive noises that are masked during the day by the presence of other, more potent stimuli.

These factors are important determinants, or regulators, of whether a particular stimulus will be attended to and perceived by an individual. The final determinant is the effect of an individual's set. There are certain conditions when the field of focus becomes quite narrowed, while a desired object or stimulus is sought in the environment. This sought-after object assumes dominance over all other perceptions. Suppose that a person is leaving for work and discovers that the car keys have been misplaced. Many objects will enter the visual field, but will not be attended to except in terms of likeness to keys, or nonlikeness to keys. If a key is found, the set is then to evaluate the likeness of that key to the one that is being sought. This evaluation relates to the specificity of the set. Attendance to the elements of the environment becomes determined exclusively by the sought-after key and the degree of similarity of objects to keys will influence the amount of attention they receive.

Factors that Exclude Stimuli from Perception

The primary regulatory mechanism that alerts the cortex to incoming stimuli can also serve to prevent the perception of stimuli. The ARAS has the dual ability to both facilitate and inhibit sensory perception. It can also function to dampen the perception of an intense stimulus to more physiologically tolerable levels. Hernandez-Peon and his associates demonstrated the process of reticular inhibition of incoming messages by inserting electrodes in the reticular formation and cochlear nucleus of a cat.[16] A repetitive tone was presented which elicited a consistent neural response. While the tone was held constant, a jar containing a white mouse was suspended in front of the cat, directly within his visual field. The cat's attention shifted to inspection of the mouse, and simultaneously the amplitude of the neural response in the cochlear nucleus decreased markedly, even though the physical stimulus remained constant. While all other conditions were held constant, the response to the tone returned to previous levels only when the mouse was removed from the visual field and the cat's attention was no longer diverted. The same mechanism operates, for example, when a person concentrates on examining a picture and does not hear a question asked of him. The ARAS functions have facilitated attention to the visual modality and simultaneously have inhibited the perception of auditory stimuli.

A second form of inhibition is related to the ability of receptors to adapt to continued stimulation. Certain types of receptors rapidly adapt to a stimulus—the rate of discharge of impulses drops even though the stimulus intensity remains constant. Touch and pressure receptors are examples of rapid adapters. When a person picks up an object, his awareness of its weight or pressure against the skin is a relatively limited experience, except under conditions of excessive pressure where conscious awareness becomes focused upon that quality. Slowly adapting receptors include pain receptors, kinesthetic receptors, and carotid sinus baroceptors.

*Factors Related to the Interpretation or
Meaning of Perceptual Contents*

Once conscious awareness of incoming sensory information has occurred, it is necessary that the information be evaluated and interpreted as to its importance or meaning to the individual. This evaluation is based upon past experience, the cultural frame of reference, the cognitive and emotional appraisal of the situation, and the set of the individual. Whereas the factors governing the transformation of the physical stimulus into a neural event and those underlying the process of attention to, or inhibition of, the incoming sensation are almost exclusively physiological in nature, the factors under consideration at this time are primarily psychological and sociocultural.

The greater the exposure of an individual to a wide variety of sounds, symbols, or situations, the greater will be the meaning acquired by these events. The more familiar a person is with a variety of perceptions, the more rapid will be his organization of current perceptions. For example, the first time a student nurse must assess the clinical status of a patient who has just experienced a myocardial infarction, she will take longer than after she has been exposed to a variety of similar situations. The ability to organize the information that needs to be assessed, to relate it to what is being perceived, and to explain it can only come about through repeated experience.

The values placed by the sociocultural regulators upon objects, relationships, and events also assist in the interpretation of the perception. When one sees a baby for the first time, the natural inclination is to relate its appearance to that of its parents. Thus the perception of the infant is not as it relates to itself, but rather as to how the hair coloring is the same as the father's and the eyes are like the mother's. The same process holds true whether the child is biologically related to the parents or has been adopted. This process is an extension of the sociocultural value that children are expected to look like its parents and dramatically influences the perception of the individual. Sociocultural values also influence the emotional valuation of the situation and may determine the "goodness" or "badness" of the perception to an individual.

The cognitive appraisal of a stimulus results in the formal labeling or interpretation of the stimulus arrived at through formal operations of logic. The work of Piaget has focused on the description of the critical stages in the development of cognitive processes.[17] Adult cognition is characterized by the ability to perform hypothetico-deductive logic and combinatorial operations in the analysis of abstract ideas. Cognitive development is sequential and orderly and can be described in terms of stages, each of which establishes a different cognitive relationship between the individual and his world. Therefore a child will interpret the meaning of a perception in a different way at various stages of his development.

Cognitive appraisal requires the solution to a problem, the problem being, "What is that stimulus which I have perceived and what does it mean to me?" The society in which one lives provides the labels by which stimuli are identified, but only the individual can add his personal meaning to the stimulus. A major theory of cognition suggests that a person formulates a hypothesis of the likeliest meaning of the stimulus based on past experience with similar stimuli and then proceeds to test the validity by responding on the basis of that hypothesis.[18] The environment either confirms or negates the initial hypothesis. If the environment fails to support it then the next likely hypothesis is tested, and so on, until the environment provides confirmation that the meaning has been identified.

Suppose you are driving down a street and in the distance you perceive a car stopped in the middle of the road. The first hypothesis might be that it has stopped for a red signal, but after searching the environment you fail to find a signal. The next hypothesis might be that the car has failed mechanically. This may or may not be substantiated. The process will continue until the appropriate interpretation of the situation, appropriate both in terms of the external situation and the internal perception of the situation, is attained.

The more unique the situation, the longer this process will require. The initial hypothesis tested is the one which the individual feels has the best chance of fitting the perception. This estimate is based upon judgment of the critical features of the situation and past interpretations of similar experiences. The principle of *generalization* provides that a person need not experience every possible situation in a class of situations in order to reach a valid interpretation. This principle indicates that an individual can respond to a wide variety of situations based on limited experience, so long as the subsequent situations retain enough aspects of the original.

Should an individual be placed in a totally unique situation where even sociocultural labels are unavailable, he does not have a pool of possible hypotheses to draw on. At such times a process of trial and error proceeds until the most likely label is determined, following which the process of sampling hypotheses can begin. A person traveling in a foreign country where he does not understand the language may find himself in a situation where sociocultural labels for events and objects are unavailable. Such situations result in hesitant, uncertain behavior reflecting the underlying state of uncertainty as to the meaning of the perceptual event.

Much of cognition involves language processing and mathematical analysis, both of which involve the use of symbols and evaluation of temporal-spatial characteristics of the stimulus as it is embedded in the environment. Studies of split-brain patients following section of the corpus callosum, which creates a condition whereby the left and right hemispheres function independently of one another, have demonstrated that these cognitive tasks are controlled by different hemispheres. Prior to these studies it had been accepted that cognition was controlled by the dominant hemisphere, i.e., the hemisphere which is the opposite of preferred handedness. If a person were right-handed, this would indicate left hemisphere dominance; left-handedness is produced by right hemisphere dominance. The assumption was that language and other cognitive processes were a result of activity of the dominant hemisphere, and that any injury to the primary centers, such as a cerebral vascular accident or extirpation, would result in loss of the cognitive ability related to that area. However, the opposite hemisphere, since it was an undeveloped source of the same potential abilities, could theoretically be trained to take over the responsibility for processing of information. The evidence now available completely revolutionizes theories of the physiological basis of cognition. The left hemisphere has been shown to function primarily in the tasks of speech, writing, and mathematics but to be virtually nonfunctional in respect to spatial relationships.[19] The right hemisphere is essentially nonverbal. It can use only a few words, perform simple addition up to ten, but can easily perform tasks involving complex cognition of spatial relationships that do not require the use of language. Gazzinaga postulated that the language of the right hemisphere might be a primitive form shared by primates.[20] It has been demonstated to be functional even if the left hemisphere language of speech is blocked by anesthesia. Filbey and Gazzinaga showed that the reaction time for verbal information (left hemisphere activity) presented to the nonverbal right hemisphere was longer than the reaction time for a nonverbal stimulus presented to the nonverbal hemisphere.[21] For the normal, intact individual, these processes form a united gestalt and cannot be divided into the components of verbal and nonverbal cognitive tasks. All evidence currently available supports the idea that the interconnection of the two hemispheres by the corpus callosum is the basis for the experience of conscious unity.

The emotional appraisal of the stimulus—the gut feeling produced by the perception—contributes to the overall meaning placed upon the event. The mood, or general affective state, of the individual will influence his interpretation of a particular perception. Moods represent a state of internal readiness for a specific emotional response. A person who is in a pleasant mood will be predisposed to attach pleasant meanings to events that coincide with that mood, whereas someone who has awakened with a short temper and mood of general irritability will be predisposed to opposite affective responses.

The process of emotional appraisal is also related to physiological changes initiated by the apperception of a stimulus, but the direct nature of the relationship is unclear. James and Lange postulated that bodily changes mediated by the autonomic nervous system following perception of an exciting or stressful stimulus constitute the emotional coloring attached to that stimulus.[22, 23] James proposed that the object stimulated the receptors and the discharge passed to the cortex where perception occurs, producing an *object-simply-apprehended*. This apprehension initiates viscertal and circulatory changes, the perception of which transforms the *object-simply-apprehended* into the *object-emotionally-felt*. Lange made a similar proposal but limited the source of the significant physiological changes to the circulatory system.

Cannon disagreed with the explanation of James and Lange and proposed that the emotional appraisal was the result of thalamic activity.[24] He found that it was possible to create a state of sham rage in cats by removal of the cortex and thalamus. This behavior of rage contained all the motor components of anger but failed to have the emotional quality, which is why the action was qualified by the term "sham." Bard later demonstrated that it was the hypothalamus, and not the thalamus, that was the source of the emotional appraisal initiated by perception, and that the visceral and circulatory changes that occur are a consequence of hypothalamic activity.[25]

Schacter and Singer explored the relationship between emotional, cognitive, and social factors in the appraisal of perceived physiological changes.[26] They induced a state of physiological activation through continuous perfusion of epinephrine, thereby creating a pattern of physiological preparedness associated with the perception of potential stress. Their results suggest that the individual's cognition of his environment constituted the basis for interpreting the physiological changes that were produced by the drug. Some subjects were placed in a rage-producing situation where the experimenter purposely became quite abusive and hostile toward them regarding their degree of cooperation.

The subjects in the "rage" situation experienced anger after the administration of epinephrine, while others who were placed in an "anxiety" situation interpreted the same physiological changes to be those related to a state of anxiety.

All these factors provide an individual with information concerning the meaning that the perception has for him. The ability to label and interpret provides structure to the environment and helps a person select the response best suited to the situation. Therefore the appraisal factors that are related to the set of the individual most directly affect the choices made within the behavior system.

Other Factors Related to Perception

Two aspects of the perceptual process remain to be discussed. The first relates to the general perceptual coping style of the individual, and the second explores the question of whether the physiological responses to stimuli are specific to the stimulus or specific to the individual.

Coping styles are general sets, or predispositions, which influence the individual's expectations concerning future events. Petrie suggested that there are two types of individuals, who modulate sensory input in different ways.[27] The first type of individual, the *augmentor*, demonstrates a tendency to augment or increase the perceived intensity of the stimulus. Such a person, hearing a sound, will report it to be louder than will others who have heard the same sound. The second type of individual, the *reducer*, tends to decrease the perceived intensity of the stimulus. Between these two extremes are the majority of individuals, who perceive the intensity of a stimulus with internal accuracy and neither inflate nor deflate their perception.

Interpersonal coping includes the ability to correctly assess the behavior of others while at the same time maintaining an integrated knowledge of oneself. Roessler has termed this the ability of the ego to maintain integrity.[28] His study of the physiological correlates of coping with stressful stimuli and their relationship to ego strength indicates that a person who has high ego strength, as measured by the scale of that name on the Minnesota Multiphasic Personality Inventory (MMPI), is more physiologically responsive to perceived threat than low-ego-strength individuals, who tend to employ defensive modes that are diffuse, nonspecific, and fail to deal efficiently with the perception of the situation.

A third approach to the study of perceptual coping styles has been proposed by Goldstein; it is related to a person's interpretation of the probability that the perceived stimulus situation will be harmful or dangerous in the future.[29, 30] *Avoiders* are individuals who hold benign expectations concerning the future events; *vigilant defenders* have threatening expectations; and the third group, *nonspecific defenders*, are those who utilize both forms of coping mechanisms, depending on the situation and their perceived ability to control it. DeLong studied the effect of information and recovery from surgery in terms of these three coping styles and found that avoiders typically had slow, complicated recoveries, and that those avoiders who were provided with specific, detailed information concerning how they would feel after their operation had more complaints of pain.[31] Vigilant defenders, or copers as termed by DeLong, recovered better when provided with specific detailed information concerning the surgery than when given only general information. The specific information apparently assisted the coper in interpreting and labeling her surgical experience (the patients were all females scheduled for abdominal hysterectomies), whereas the avoider was unable to adequately defend herself against the experience when provided with specific information.

The last area of concern related to the perceptual process is the relationship between the characteristics of the stimulus and the physiological changes which accompany perception. Maintenance of a state of physiological equilibrium, or homeostasis, is mediated primarily through the two branches of the *autonomic nervous system* (ANS): the *sympathetic nervous system* (SNS) and the *parasympathetic nervous system* (PNS). Parasympathetic activity is predominantly associated with conservation of energy and life-maintenance functions. For example, digestion, gastrointestinal function, and sleep are facilitated by the PNS. The SNS, in contrast, prepares an individual for a response to an emergency. This coordinated pattern of physiological activation is the *fight or flight* response described by Cannon and is initiated by the perception of an unusual or threatening stimulus in the environment.[32] After the stimulus has been labeled and interpreted, this nonspecific pattern of activation is altered. The problem is whether the subsequent pattern of physiological activation is specific to the stimulus or to the individual.

Wenger proposed that the physiological response to a stimulus is *specific to the properties of the stimulus and to the emotional label* attached to the perception.[33] The pattern of response to the class of stimuli would be constant, but the magnitude of the response would vary based on the intensity of the stimulus and the level of physiological activity prior to the onset of the perceptual process. If an individual saw two people, one of whom she loved and the other one with whom she had a friendly relationship, the same pattern of physiological responses would occur, but in the former case, the changes would be of greater magnitude. Based on this assumption of stimulus specificity, it would be possible to compare the characteristics of individuals and their magnitude of response to various stimuli.

Lacey proposed the opposite position, that an individual responds to all incoming stimuli with a similar physiological pattern that is *specific to the individual* and not to the stimulus properties.[34] His hypothesis of response-specificity states that regardless of the type of stimulus presented, an individual will respond with maximal activation in the same physiological function; the activation being characteristic of the individual, not of the stimulus.

Evidence indicates that what occurs when an individual perceives a stimulus is determined both by the stimulus and by the individual. Engel demonstrated that physiological responses to different stimuli are related to the stimulus eliciting the response, but that within that specific pattern the individual may respond with the maximal change in an idiosyncratic manner.[35] One person may respond maximally with a change in heart rate, a second person with changes in the gastrointestinal tract, and a third with respiratory alterations, but all three will show the *same overall pattern* of physiological change to the identical stimulus event.

SUMMARY. The result of all these factors related to perception, labeling, and interpretation of environmental stimuli constitutes the information utilized by the behavioral system for selecting a response to the environment. The information is based upon the complex analysis of the stimulus by the physiological, psychological, and sociocultural factors relevant to an individual. The speed with which this analytical process takes place is such that it is impossible to separate the whole into the individual components while the experience is occurring. Illness, extreme stress, aging, sensory deficits, and changes in sociocultural conditions will affect the process and may change the form of behavioral response or cause a delay between the stimulus event and the response.

Once the perception has been interpreted, additional factors determine the form of the subsequent response. These factors can be general to all eight subsytems of behavior or specific to one subsystem. Chapter 6 will discuss the general regulators of subsystem function.

STUDY QUESTIONS

1. Describe the critical sensory dimensions contained in a hospital room, including objects, persons, and components of the physical environment.

2. How does the cognitive interpretation of hospitalization affect an individual's perception of the situation around him?

3. What physical stimuli, chemical stimuli, and organic factors can affect attention?

4. After you describe the sensory dimensions contained in the hospital room, describe the same environment while:
   a. blindfolded
   b. wearing ear plugs

  c. simultaneously blindfolded and wearing ear plugs

5. How does the sociocultural environment influence the perception of a stimulus? Select a cultural system other than your own and compare the two for similarities and dissimilarities in the perception of a cultural event such as death.

six | General Regulators of Behavioral System Activity

The dimensions of the individual that influence behavioral system activity and response patterns are extensive. Chapter 5 discussed the many variables associated with perception of the external environment; the end result of the perceptual process comprises the immediate input for behavioral system function. The following discussion covers some factors that regulate or influence behavioral responses, including (1) genetic inheritance, e.g., gender determinants and intelligence; (2) circadian rhythms; (3) age; (4) sex; (5) attitudes, values, and beliefs; (6) social class and role expectations; (7) locus of control and alienation; (8) creativity and problem-solving ability; and (9) self-concept. These general regulators may be viewed as a class of intervening variables in that they influence the internalized perception of a situation as well as the activity of the structural units of some or all of the subsystems of behavior, i.e., goal, choice, set, and acts. Their primary effects are to limit and direct the form of information input made available to the behavioral system as a whole and to significantly determine the subsequent behavioral response pattern utilized by the individual.

116

GENETIC INHERITANCE

At the moment of conception an individual is endowed with those factors of his ancestors that have been transmitted from generation to generation by the genetic material. The nature of the influence of genetic factors has been most clearly identified in relation to physical characteristics, sex determinants, and intelligence, although it is reasonable to assume that the impact of inheritance is evident in all activities of the behavioral system. The nature of the interaction between factors of inheritance that exist *in potentia* at birth and factors of the environment in which the development of the individual takes place has been a topic of controversy for decades. The primary question has been whether the inherited aspects of the individual represent the upper limit of potential development, or whether the impact of the environment can result in a significant expansion of native abilities. Studies of identical twins who were separated shortly after birth and raised in different environments have contributed the greatest amount of information concerning this question. The assumption underlying this approach has been that those behaviors that are basically determined by genetic factors will demonstrate a high correlation in the separated twins whereas those that are environmentally determined will demonstrate little or no correlation.

Physical characteristics of individuals include such factors as sex, skin color, eye color, hair color and degree of curl, facial structure, and body height. The distinctiveness of inherited characteristics among brothers and sisters of the same parents illustrates the extensive variation of possible genetic combinations. Probability studies have estimated that a male can produce sperm of 8,388,608 genetically different combinations, and a female can produce a like variety of ova.[1] Each of the 8 million genetic combinations of the male has an equal chance of uniting with any one of the 8 million genetic combinations of the female, resulting in an almost infinite number of possible combinations. Thus, it is not hard to understand why a couple may have, for example, three children, each of whom differs from the others in physical and temperamental characteristics.

The human cell has twenty-three pairs of chromosomes, twenty-two pairs of which are termed *autosomes*. Autosomes are responsible for transmission of genetic characteristics that are inherited under the Mendelian principle of dominant and recessive traits, irrespective of the sex of the parent from which they originated. The twenty-third pair, the sex chromosome pair, determines the sex of the individual. The X chromosome carries a full complement of genes, whereas the Y chromosome is almost totally deficient in genetic material but serves the function of determining maleness. If the union of sperm and ovum results in an XY pair the child will be male; an XX pair will produce a female. Because of the lack of genetic material in the Y chromosome, any recessive trait carried by the X chromosome of the mother will be fully expressed in the male child, whereas recessive traits in the female will be expressed only if the paternal X chromosome also contains the recessive trait. This explains why sex-linked recessive traits such as color blindness and hemophilia occur with a much higher frequency among male children.[2] The relationship of gender and behavior will be discussed in a separate section, because of the widespread interaction between physical sex and sociocultural factors and their influence on the behavioral response patterns of all eight subsystems.

The physical characteristics a child is born with influence the environmental response to the child, and his response to his environmental perceptions. Those individuals who inherit such recessive traits as deaf-mutism and cleft-palate will be required to develop compensatory modes of ingestive behaviors. It is likely that the environmental response toward such children will be different from that which is extended to the normal, healthy child. The archetypal infant in the American culture is a healthy, blue-eyed, blonde-haired boy—see any baby food advertisement. Infants born with these characteristics tend to receive a greater positive response from adults than do infants with dominant traits such as dark hair and brown eyes. It may be that because these latter characteristics are more common in the population, they elicit a different response. It is not known whether such early experiences affect the later development of affiliative relationships with others, but it does seem possible that an infant who receives a high amount of positive reinforcement and interaction from the environment for his physical characteristics may develop a positive self-concept and a sense of distinctiveness in relation to others.

There are some indirect indications that there is a genetic basis to temperamental characteristics. Observations of neonate behavior tend to support an inherited contribution. Some infants show a placid response to events in the environment, while others are hypersensitive to stimuli and have a high level of activity. Mothers often report that whereas one infant loved to be held and cuddled, a second child refused all attempts to be held and became extremely irritable when cuddled. These early responses affect the amount of exposure an infant will have with stimuli associated with cuddling, such as the body warmth of the mother, movement, and the attention and interaction associated with such events. Whether or not such experiences affect mature behavior is not known at this time, but it would again seem reasonable to assume at least an indirect effect upon the development of affiliative and dependency behavioral patterns.

The study of the genetic basis of personality traits is relatively new. There appears to be evidence to support an inherited predisposition to the development of schizophrenia and affective mood disorders such as manic-depressive psychosis. Studies of normal development of personality characteristics are few. Juel-Neilsen studied the personality and social behaviors of twelve sets of twins who had been reared apart.[3] He reported that the twins were similar in terms of expressive movements such as gestures, and in types of complaints of psychosomatic origin. They were dissimilar in terms of socially based behaviors such as personal values, choice of marital partner, style of interpersonal interaction, fields of interest, level of ambition, style of dress, and level of aggressiveness. He concluded that the personalities of the twins showed a "moderate" degree of resemblance. Other studies involving identical twins and their personality profiles indicate that there is a significant degree of similarity of interests and personality characteristics,[4] but whether this is because the environment tends to reinforce and encourage the development of similar interests in identical twins or because it is a genetically determined similarity is still open to question.

Intelligence is an abstract concept that is related to an individual's ability to deal with complex, abstract material. It has been estimated by Burt that 77 percent of an individual's performance on an intelligence test is related to genetic factors; the remaining 23 percent represents sociocultural and environmental influences.[5] The activity of the achievement subsystem of behavior, which is related to the ability of an individual to master himself and his environment, is directly regulated to a significant degree by level of intelligence. Studies have demonstrated a positive relationship between intelligence quotient scores (IQ) and occupation, socioeconomic status, academic grades, aptitudes, and performance variables. Not all these relationships are independent of one another. For example, occupations such as medicine and law that require the ability to achieve high academic marks also tend to place the individual in a middle upper, or upper socioeconomic class.

Measurement of intelligence levels is through the application of standardized tests such as the Stanford-Binet or the Wechsler Adult Intelligence Scale (WAIS). The average or mean intelligence quotient of the population is arbitrarily set at 100. Persons with an IQ of 120 or above are expected to have little difficulty with academic performance in a university setting, given the motivation to succeed. Mental retardation is defined as an IQ of 70 or below.

Many factors determine an individual's test score, including the level of motivation at the time of testing; the particular meaning of the measurement to the individual; the level of test anxiety; and whether the test items relate to information contained within the ethnic background of the child. This last factor is a major reason why children of lower socioeconomic groups and minority ethnic groups have consistently shown lower intelligence scores, compared to children of the majority ethnic group. Children reared in sociocultural settings other than the one upon which the test was constructed are at a disadvantage, since what the test essentially is measuring is the amount of exposure of the child to commonly shared knowledge. Such government programs as Head Start, and television programs as "Sesame Street" and "Electric Company" have been developed for children of relatively impoverished environments in order to decrease the disparity in exposure to sociocultural knowledge in minority ethnic groups. The contention was that exposure to this common knowledge of the majority culture before entry into school would enable these children to achieve greater success in intellectual tasks at an early age. Studies of academic performance of "Sesame Street graduates" have not consistently supported this contention; instead some results have indicated that these children experience *greater* difficulties and perform more poorly in the first grade than a group of similar children who have not been exposed to "Sesame Street."[6] In contrast, children who attented structured Head Start programs demonstrated significantly more educational gains in grade school performance.

A second factor that may result in depressed IQ scores is related to the quality of the educational system to which the child is exposed. IQ measurements were obtained for a group of black children in the southern United States. These same children later moved to the North and subsequent measurements demonstrated a significant increase in IQ. It was felt that the increase in scores was greater than would be expected if it were a result of age. In general, IQ scores remain fairly stable throughout a lifetime. The most rapid changes in intelligence occur up to age 16, with slow, upward movement in intelligence scores until the ages of 50 to 60. However, IQ scores control for these expected changes in intelligence, so that the scores themselves should remain relatively stable. Given this stability it has been demonstrated that a child's intelligence quotient from age six on is highly correlated with achievement behavior and intellectual mastery during adolescence and adulthood.

The exact abilities that relate to intelligence have not been agreed upon. Thurstone proposed seven primary mental abilities based upon a factor analysis of a variety of test scores.[7] These seven primary abilities are: (1) number; (2) word fluency; (3) verbal meaning; (4) memory; (5) reasoning; (6) spatial perception; and (7) perceptual speed. Number refers to the ability to perform mathematical operations such as addition, subtraction, multiplication, and division. Word fluency involves the ease of speaking and writing, whereas verbal meaning involves the comprehension of ideas in word symbols. Memory involves the ability to retain and revive impressions, or the recall of past experiences. Reasoning is the ability to solve complex problems and make predictions based upon past experience; it is also related to a creative approach to one's environment. Spatial perception and perceptual speed relate to informational aspects of the environment, the spatial factor being the ability to perceive size and spatial relationships, and the perceptual speed a measure of the ability to identify stimulus objects quickly and accurately. These abilities are believed to be innate characteristics of the individual, primarily determined by genetic factors rather than learned behavior, which is acquired through environmental interaction.

As our knowledge of the relationship between behavior and inheritance expands, the significance of the moment of conception and its lasting influence upon the individual will be more deeply appreciated. For the most part these genetic factors are beyond the immediate control of the environment and of the individual, although they are realized to their greatest extent only with the help of a rich and meaningful environment.

CIRCADIAN AND PSYCHOBIOLOGICAL RHYTHMS

Circadian and other psychobiological rhythms represent a complex interdependence between the temporal-spatial qualities of the physical environment and the internal processes of the organism. Any open system, such as the human being, will demonstrate a correlation of psychophysiological function with the environment. Circadian rhythms specifically represent those systematic cycles that have a periodicity of approximately 24 hours (*circa*—about, *dian*—day). A number of other psychophysiological rhythms of shorter or longer duration have also been identified. The menstrual cycle, for example, is a biological rhythm with a periodicity of approximately 28 days, plus or minus 6 days.

Circadian rhythms are basically sinusoidal, with a peak point of function, followed by a trough or low point in activity, the times between peak and trough, and trough and succeeding peak being equal. Among the physiological functions that have been demonstrated to follow a circadian pattern are: body temperature[8]; plasma levels of 17-hydroxycorticosteroids; sleep-wakefulness cycles; skin resistance; urinary excretion of sodium and potassium; water excretion; heart rate; fluctuations in blood eosinophils; serum iron content; blood pressure; pituitary release of adrenocorticotrophic hormone; and blood sugar levels. At birth these functions show a constant level of function over 24 hours; each psysiological variable becomes cyclic at different stages of early development.[9]

Under natural environmental conditions the circadian rhythms are entrained, i.e., synchronized, by an environmental factor called the *zeitgeber* (*zeit*—time; *geber*—giver). One of the most important zeitgebers for animals is the 24-hour cycle of light and dark, but the importance of light for human beings is questionable since most persons reside in settings that contain artificial sources of light. A study by Aschoff, Fatranská, and Giedke demonstrated that knowledge of time of day, living routine, and social communication are major zeitgebers for the human circadian rhythms, and that social cues are "sufficient" to entrain human circadian rhythms.[10]

Lobban attempted to dissociate the normal circadian patterns of body temperature and urinary secretion of sodium, potassium, chloride, and water (volume) from real time by giving watches to people that showed elapsed times of 24 hours, whereas the real time was either 22, 24, or 27 hours. [11] She was able to demonstrate that the excretion of water followed the new elapsed time whereas the other functions remained on a fixed 24-hour cycle. The impact of time is most readily apparent on activity levels, sleep-wakefulness cycles, and body temperature curves when a person travels to a new time zone. McFarland suggests that it requires approximately two and one-half days of rest to adjust to a time zone change of 4–6 hours.[12] During this adjustment of activity patterns and sleep patterns to the new time, individuals often experience changes in moods and tolerance for new experiences and have a lessened ability to make decisions.

Richter presented evidence for biological clocks in psychological processes.[13] Alternating periods of mania and depression may follow the principles of a biological rhythm. He reported a case of a woman with Parkinsonism whose symptoms varied on a 24-hour cycle. Between the hours of noon and 9:00 P.M. she experienced marked rigidity accompanied by mental alertness; while between the hours of 9:00 P.M. and noon the physical symptoms improved markedly, but the improvement was accompanied by depressive feelings. He suggests that disorders such as peptic ulcer, schizophrenia, hydroarthosis, epilepsy, and depression also demonstrate underlying rhythmic patterns of varying lengths of time.

These circadian and biological rhythms serve a general organizing function of the behavioral system. Analysis of input from the environment is facilitated by optimum function. This suggests that the importance of the relationship between time of day and the efficiency of the behavioral system must be considered. If a response is required of the system at a time of minimal function, there will be a greater demand for energy expenditure and a longer time for the appearance of a response, compared to conditions of maximal function. Processing of the input from the environment—perception —will be slowed, and deciding which response will be most functional and adaptive in the setting will require greater effort. Many individuals, for example, are able to function most efficiently in the morning hours when their body temperature is ascending to an early peak and most other functions are at, or near, the point of maximal efficiency. Other people are unable to function in the morning but rather are most efficient in the late afternoon or evening hours. This latter group can be shown to have their peak efficiency curves coincident with the late afternoon. If such people are required to make decisions in the morning, at a time of lowered efficiency, they will experience great difficulty, and may vacillate regarding their choice. Given the identical task in the afternoon, their behavior will not demonstrate such vacillation.

AGE

The relationship between the chronological age of an individual and behavior is an entire field of study itself. This discussion can only touch on some of the relationships of age and subsystem activity. Response patterns and capabilities are minimal in the first years of life and near the end of life, but for very different reasons. The infant has few response patterns available because of two major factors: a lack of experience with a variety of stimuli present in the environment; and a state of physiological and psychological immaturity. The aged individual, at the other extreme, has a limited range of responses available because of declining sensory perceptual monitoring of the environment and declining physical abilities. These limitations are acknowledged by our sociocultural institutions, which provide various forms of assistance to ensure the continued survival of the individual. Federal social security and Medicare programs both provide forms of economic and health assistance to older, retired members of our society, enabling them to maintain a level of financial independence. Foster homes are also supported and maintained by society to care for children who are not, or cannot be, cared for by their parents.

The period of greatest availability of responses to the requirements of the environment are the years of early to middle adulthood —the peak is attained at about age 35 to 40. This age of peak performance varies for each individual. The period of greatest creativity within occupational groups, as measured by the age at which the greatest achievement is made, varies with the degree of physical or intellectual requirements of the field. Athletes tend to reach their peak performance in their mid-twenties, whereas philosophers and physicians do not attain peak performance until 35 to 40.

An infant is totally dependent upon others in the environment to provide those elements necessary for future development and survival. Even though the human being achieves relatively greater flexibility and independence of behavior from his surrounding environment, as compared to other primates, the process requires the most extensive and prolonged period of dependency known in the animal kingdom. The ability of an infant to discriminate objects in his immediate environment is not developed in the early weeks of infancy, and during this time he will accept any object as sufficient to satisfy his needs. This period represents a preingestive period of total dependency upon the environment. Once the child develops the ability to identify his mother, or the chief figure of gratification, from others in the environment there develops a period of time during which only the maternal figure is able to gratify dependency requirements. The infant's cry is sufficient to ensure her presence, following which she must identify the source of the problem in order to produce a return to a state of equilibrium. During this period of total dependency, the adults in the environment not only respond to the infant's cries but also anticipate areas in which specific forms of assistance are needed, such as in feeding, manipulation of objects, removal of products of elimination, and so forth. In a sense the early infancy period is one in which the maternal figure serves as a discriminatory behavioral system for the infant until such time as the child begins to develop response modes.

An infant is exposed to a wide variety of potential threats within his environment from which he must be protected so that he can survive. Because of the prepotency of biological factors for survival, threats to continued biological existence are responded to with some form of behavior, whereas social and psychological threats may not be directly registered and may not be responded to in any manner. The responses available to the infant are reflexive and global in nature, and are characterized as *nonspecific protective behaviors*. The ability to cry in order to summon assistance is the only response available to the infant when a threat that cannot be overcome solely through a primitive reflexive maneuver is present in the environment.

Ingestive, eliminative, and restorative behaviors are primarily determined by physiological functions in the infant. These behaviors dominate the 24-hour cycle, and are expressed in the feeling that all an infant seems to do is sleep, eat, and eliminate at varying intervals. Infants experience themselves as part of their environment; in other words, the infant does not "know" that he is separate from his surroundings.

As a child matures, an element of intention and purposiveness begins to characterize those behavioral patterns previously performed on an involuntary basis. The primary requirement for the development of voluntary action is physiological maturation. Control of anal and urethral sphincters for the elimination of fecal and urinary waste products must be feasible on a physiological level before the sociocultural regulator of toilet training can be instituted as an expectation of behavior. The development of the ability to move from one environment to another and the rapidly emerging ability to express internal needs in verbal language greatly expedite the relationship of the child to his environment. The general characteristic of early childhood behaviors is a rapid increase in available responses to the environment, responses that vary in terms of complexity and appropriateness, but that serve to bring meaning to external events. As the responses increase in availability, a child begins to be exposed to a greater variety of environments in which elementary discriminations are made. Sociocultural regulators such as attitudes, values, and role expectations begin to exert an ever-increasing influence upon behavior. During early childhood these sociocultural expectations tend to be applied benevolently, in that transgressions or failures are expected and are rarely punished; this is not the case in later childhood. The toddler who has been fully toilet trained is rarely punished for an accident, but the older the child becomes, the more exacting are the societal requirements. In the Japanese culture, a child is traditionally allowed total freedom of expression until the age of seven at which time he is expected to assume the restraint characteristic of an adult.

The preadolescent years reflect a movement into the areas of developing intellectual learning, increased fine motor development, and a repertoire of achievement behaviors. The tasks that confront a child are complex and require a long period of development before they are internally available for response with required behavioral patterns. The tasks associated with the eight subsystems vary in terms of complexity. Ingestive behaviors in relation to food and sensory stimuli are fairly well established, whereas the establishment of reading skills and the acquisition of knowledge are emerging behaviors of importance. As the number of persons with whom a child comes into contact increases, he devotes more and more attention to the development of successful interpersonal behaviors with friends and peers. The recognition of a need for other people simply for the pleasure of interaction (e.g., affiliative behaviors) becomes more and more important in the child's pattern of activity. The parents begin to be replaced by the teacher and other adults as the "experts" in the environment. The privilege of the parent to continue to make decisions and provide assistance is constantly challenged by the child's peer group.

By adolescence, the sociocultural regulators have assumed a primary level of significance in the performance and selection of behaviors. Many behavioral patterns of adulthood have been sufficiently established by this time so as to be readily identified by others. The major difference between adolescence and adulthood is that society does not require the adolescent to assume full responsibility for his actions, although he will be held increasingly accountable for his behavior. The impetus is toward increasing self-reliance and exposure to a greater variety of experiences in order to develop an extensive repertoire of behavioral responses.

Adulthood is an extended phase during which an individual is expected to demonstrate self-reliance in tasks related to himself, as well as to function as an effective contributor to society. Illness is one of the few conditions that is accepted as a valid situation for dependency on others; employment of persons who are skilled in trades other than that which the individual has developed for himself is also acceptable. Achievement behaviors are dominant in adulthood, as are sexual and affiliative response patterns, since the basic task of this period of life is to establish one's own relationship to the environment and to provide new members of the group through parenthood.

The final phase of old age reflects a period of increasing dependency upon others for completion of tasks in which independence was previously possible. As a person ages, physical abilities become less precise, sensory modalities may develop deficits which result in decreased ingestion and perception of the environment, and the individual once again must rely on persons and objects in the environment to provide for his continued existence. Shakespeare's description of the seven ages of man in *As You Like It* is a fairly accurate portrayal of the developmental periods of an individual, and the inevitable decline of ability. There is a withdrawal from involvement with others, with the possible exception of family members, resulting in a decrease in affiliative behaviors. An elderly person may allow another person to care for him and do that which he can no longer do for himself, but this interaction is often on a highly formalized basis and is mediated through activity of the dependency subsystem. Earlier concerns with achievement, approval, and attention become less and less important, until for some persons even continued survival becomes a burden, no longer sought but only tolerated.

SEX

The sexual identity of an individual, determined at the time of conception and molded primarily through the influences of the sociocultural environment, significantly affects the patterns of behavioral responses developed by that individual. A major area of disagreement between society and the members of the feminist movement within the United States is whether many of the apparent differences between males and females in terms of achievement and dependency behaviors are a function of sociocultural bias, or whether there are basic biological differences that serve to produce the observed differences in behaviors between the sexes.

Cross-cultural studies demonstrate that division of labor is made on the basis of sex, although not all cultures designate the maintenance of home and care of children as the primary areas of female responsibility. Dornbusch suggested that the function of distinguishing cultural expectations for individuals on the basis of sex is that it serves as a justification for not exposing all members of a cultural group to all potential activities.[14] If society has clear expectations that girls will engage in one limited set of behavioral activities and boys will engage in other activities, it should be more advantageous to structure the exposure of each sex to these limited areas.

The impact of differential approaches to masculine and feminine sex roles is established before birth. Often the assumption is made that a couple would prefer the birth of a boy, as though the attribute of maleness is preferable to that of femaleness. In many hospitals the discrimination of ascribed sexual identity begins with wrapping an infant in a blue or pink blanket. From that moment on, significant distinctions in the expected and valued behaviors for each sex are provided continuously by the environment. For example, girls tend to be reinforced for relative passivity and dependence upon others in a threatening situation, but boys are admonished to face up to life "like a man." The implication is that to be "like a man" is to actively attack the perceived source of threat, be it person, object, or idea. These early child-rearing differences result in the encouragement of very different behaviors that become sex-linked by virtue of repetitive reinforcement.

With the possible exception of the ingestive behavioral subsystem the response patterns of the seven other subsystems may be greatly modified by the sex of the individual. One of the first distinctions is presented by the approach to toilet training. The sociocultural distinction of boys' standing up to urinate and girls' sitting down has an anatomical basis, because of the absence of a penis in the female requires the sitting position for ease of sphincter relaxation.

The cultural values applied to the expression of feelings appear to be a major determinant of differences in response patterns between males and females. Women tend to be reinforced for expressing how they feel, with the possible exception of anger. It is through the expression of feelings that dependency is encouraged and reliance upon others for emotional support develops. Men, on the other hand, tend to be reinforced for keeping their feelings inside, for not allowing emotions to enter into decision making. Women who behave this way may be described as "castrating" or "masculine." Aggressiveness, independence, objectivity, and lack of emotional expression either through denial or repression are all valued "masculine" behaviors in the American culture. Women, in contrast, are described as "feminine" if they are passive, dependent, emotional, expressive, and subjective in their behavioral responses.

Maccoby and Jacklin recently examined all the stereotypic statements concerning differences between boys and girls by reviewing the scientific literature for evidence in scientific studies in support of, or disagreement with, the cultural expectations.[15] They concluded that the evidence was not available to support many stereotypes of sexual differences and suggested that the reason the culture continues to foster these beliefs is a result of selective attention. If a member of the sex behaves as expected, an observer will note this as confirmation of his expectation; if a member of the sex in question does not behave in accordance with expectations, the behavior is unlikely to be noticed or it will be rationalized as an unusual occurrence. In other words, these stereotypes are maintained through selective attention to data that conform to expectations.

In the sexual subsystem of behavior, there are biological and sociocultural determinants of response patterns. The experience of pregnancy and childbirth is an exclusively feminine occurrence. The man is a highly important contributor to the development of this event through his impregnation of the woman, but beyond that, any knowledge of the components of the experience of pregnancy and birth can be acquired only vicariously. Perhaps it is because of this distinctive function of the female in ensuring the continuance of the sociocultural group that socialization for girls is directed primarily toward the "feminine" roles of growing up, getting married, and having children, as opposed to the "masculine" roles of growing up, studying hard, and earning a high salary. Society in the United States has tended to equate feminine achievement with behaviors related to the sexual subsystem of behavior. This equation is undergoing change as the upper limit of acceptable population density is neared. Women are no longer considered deviant if they fail to marry or elect not to bear children. Rather the sociocultural values are beginning to emphasize the value of not bearing children as the concern with shortages of food and resources for survival of those persons already existing becomes more and more acute.

The sociocultural regulators that influence the development of affiliative behaviors and roles also demonstrate a differentiation in terms of sexual factors. The role expectations associated with mothering are, in our society, fulfilled by females, whereas the expectations of fathering are fulfilled by males. A child initially relates to qualities of maleness and femaleness in toys and play through their identification of "mother" and "father." Boys are encouraged to develop rough, manly styles of interaction, whereas girls are encouraged to play with dolls and develop the feminine role behaviors required for adult life. As a result of the different style of interaction, affiliative behaviors tend to be restricted to individuals of the same sex throughout childhod and adolescence. In the adult years, it is still a sociocultural expectation that men will engage in activities and interactions with other men as exemplified by the tradition of the men retiring to a room for a drink and male companionship following a meal, or of a night out with "the boys." Similar expectations apply to women. A husband who elects to affiliate primarily with his wife runs the danger of being identified as "henpecked."

The sex of an individual indirectly regulates the acceptance of varying degrees of dependency upon others. There is no biological reason for assuming that men are inherently more independent and self-reliant than women and yet, as a result of the sociocultural values, there are significant differences between males and females in terms of dependency. Girls receive reinforcement for dependency upon others, whereas boys are punished for failure to rely upon themselves. The punishment of dependent actions in boys results in high levels of anxiety when, as adults, they encounter a situation which implicitly requires dependency upon others, such as hospitalization. It is not unusual for the dominant basis of relationships with others to be determined on the basis of dependency. Many marriages are primarily relationships in which the requirements for dependency on the part of the wife, and the requirements for independence on the part of the husband, can be gratified. Such relationships require continued dominance on the part of one partner and submissiveness on the part of the other and may encounter difficulty should a change in behavior occur on the part of the submissive member. As child rearing practices become the same or similar for both sexes there should be a significant decrease in the apparent sexual differences related to dependency behaviors.

As indicated earlier, feminine achievement behaviors are mediated primarily through the goals of the sexual subsystem. If a girl does desire to enter an occupational group, once again the criterion of sex may significantly restrict her available choices. Until recently Americans have believed that women should be nurses and primary grade school teachers, and males should be doctors and college professors. Even within an occupational class there are apparent sexual differences. Administrators tend to be males, and even in fields dominated by women, e.g., nursing, men will be promoted to positions of responsibility faster than will women with comparable skills. In past years, the primary goal of achievement for a woman was attained when she married and had children, whereas for a man the primary goal of achievement was related to success within his chosen occupation. Even today, a married professional woman is expected by society to subsume her professional goals to those of her husband. She is expected to seek employment in the geographical setting that will be most favorable to her husband's achievement goals. There is still an unspoken conviction by some that the sole motivation for a woman entering college, graduate school, or a profession is to get herself a husband. This assumption is most apparent in the difficulties experienced by women seeking admission to various fields of study, and in the quota system for admission in fields such as medicine. In the past the quota system was implicitly used to keep the number of minorities and women at a minimum. Affirmative action programs now establish an explicit quota in order to promote entry of these same groups into previously closed fields of study.

Almost all studies of achievement behavior have consistently demonstrated differences between males and females, even though there are no sexual differences in intelligence scores. Horner attempted to explain the basis of these consistent differences.[16] She concluded that women appear to have the motivation to "avoid success," or the need to fail, as a means of fulfilling their perception of sociocultural expectations. Women tend to equate intellectual achievement with a loss of femininity. Therefore, a woman is confronted with a conflict situation whereby she can succeed intellectually and meet her own internal standard while failing to meet the societal expectations associated with femininity, or she can let herself fail and be rewarded by sociocultural approval. The element of competition has a negative effect on feminine achievement, whereas it is a positive reinforcer for much of male achievement behavior, such as in athletics, politics, courtroom debates, and so forth. As the sociocultural prohibitions against feminine achievement are modified it will be possible to more clearly identify those areas, if any, in which males and females truly differ.

A major physiological regulator of female behavior is the menstrual cycle and the effects of varying levels of circulating hormones upon behavior. No comparable process is known to exist in the male.

There is evidence that a greater proportion of women who commit suicide and criminal acts of violence, and who have serious or fatal plane accidents as pilots, do so during the premenstrual and menstrual phases of the cycle.[17,18] In countries with the Napolenic Code of Law, a reasonable plea of insanity can be entered for crimes committed during the premenstrual phase.[19] Benedek and Rubenstein studied the relationship between content of psychodynamic material and variations in vaginal smears throughout the menstrual cycle.[20] They reported that in the follicular phase of the cycle women tended to experience greater heterosexual tendencies, whereas during the luteal phase, following ovulation, the women were more preoccupied with fantasy material related to being taken care of, i.e., dependency subsystem goals. It might be predicted that a greater frequency of sexual subsystem behaviors would be present prior to ovulation when pregnancy is a possibility than following ovulation at which time dependency behaviors might dominate. Whether the mood fluctuations that have been described to be consistently related to the menstrual cycle[21,22] are related to hormonal changes or are learned behaviors of a sociocultural etiology is not yet known. What does seem evident is that significant life events have been shown to be related to the various phases of the menstrual cycle and as such must be considered in the understanding of behavior.

ATTITUDES, VALUES, AND BELIEFS

A major mode of transmission of sociocultural information is through the development of personal attitudes, values, and beliefs that are consonant with the values of a particular society. *Attitudes* are dispositions or feelings toward a person object, or idea; they are generally agreed to have three components: cognitive, affective, and behavioral. The cognitive component is the knowledge associated with the attitude. *Beliefs* represent personal confidence in the validity or existence of some idea, person, or object that is not immediately verifiable. Therefore beliefs are a special class of attitudes in which the cognitive component is not based on fact, but on faith. *Values* represent the affective component associated with the person, object, or idea with which the attitude is concerned. Some experts limit the concept of attitude to the cognitive and affective components and view behavior as a separate concept that may or may not be related to the attitude. For example, cigarette smokers may acknowledge that smoking is harmful to health (cognitive and affective elements), and yet they continue to smoke (behavior). If the behavioral actions are defined as part of the attitude then the above example contains elements incongruent to one another. It is possible, however, that the real attitude of the smoker is that smoking will not harm him, and he is verbalizing what seems to be the expected attitude of the group and his belief as to what might happen to others, rather than his own personal attitude. In such a situation, his personal attitude and behavior are congruent, while the collective attitude expressed in the original statement continues to be incongruent in terms of observed behavior and stated attitude.

Attitudes can be positive or negative in value and they can be characterized by five primary attributes or dimensions.[23] *Extremeness* of an attitudes relates to the relative degree of positive or negative value attached to the object. For example, attitudes toward the various periods of classical music may be positive or negative and will affect whether the behavior of listening is engaged in at a given time. A person may be favorably disposed to baroque music, moderately positive toward music of the romantic period, and intensely negative toward music of the twentieth century. It is not unusual for persons at a concert to rise and depart in the middle of a performance of a musical composition they intensely dislike as they would rise to a standing ovation for music they greatly appreciate.

The *content* of an attitude refers to the precise meaning of the event or object toward which the attitude is directed. All those people who enjoy baroque music may do so for different reasons. One may particularly appreciate the counterpoint, another the harmony, and a third the rhythm. In other words, the basic statement of liking as an attitude is a different "liking" to each person. This issue of the meaning of an attitude is one of the problems inherent in interpreting public opinion polls when questions are asked such as, "Do you approve of his performance in that position?" One's perception of the performance of the individual and of the responsibilities of the position are dimensions of the attitude that are open to a diversity of interpretations, so that one person may respond to different meanings and contents than may the remainder of the group.

The clarity, or structure, present in an attitude may be either highly differentiated or totally lacking. The degree of *differentiation* contained in an attitude is a measure of the ability of an individual to specify what aspect of the object or event results in the attitude. An individual may be able to state exactly what it is about twentieth century music that creates a negative attitude. It may be the atonality, the lack of a readily identifiable theme, the discordance aroused with himself, or any number of other reasons. This attitude clearly has a greater degree of differentiation than the attitude of a person who cannot identify what it is that he doesn't like, but just knows that he doesn't like it. The diffuseness of the latter attitude is particularly evident in stereotypic attitudes in which an individual has integrated a sociocultural attitude into his personal belief system without any conscious examination of the validity of the concept for him.

Attitudes also differ in their degree of *integration* or isolation in relation to other attitudes of the individual. A person's negative attitude toward modern music may result in, or influence, attitudes toward classical music in general, toward individuals who express a positive liking of modern music, or toward orchestra conductors who perform modern works. This single attitude may become associated with attitudes toward other areas of life that are totally unrelated to music such as politics, religion, and so forth. If the value of the attitude becomes a determinant of other attitudinal responses, it is a highly integrated attitude. It may be difficult to identify the potent precursors of a particular attitude in such a situation.

A final dimension related to attitudes is the *strength* of the belief when evidence to the contrary is presented. Will a person confronted with direct, objective evidence that would appear to totally contradict the cognitive content of the attitude still retain it? Many attitudes are strong and highly resistant to change. The stereotyping of males and females even though there is no objective evidence to support significant differences in performance is an example of a rigid, resistant attitude. If an attitude easily changes when external evidence is presented to the contrary, then that attitude is weak. Festinger suggested that when a person is experiencing *cognitive dissonance*, defined as an awareness of an incongruity between an environmental event and an individual's personal attitudes toward the event, the dissonance must be resolved in some way that is adaptive to the individual.[24] This may be done through changes in the attitude, through rationalization that the event is really different than it appears, or through justification of the incongruity. This theory emerged from observations of members of a religious sect that had predicted the end of the world. All the members had made the necessary preparations for the event to occur on the designated day. When the event failed to occur some individuals still retained their attitude toward the pending apocalypse but reduced their dissonance by assuming that it was the wrong year, or that, for whatever reason, they were required to wait for a longer time. Others changed their attitude and departed from the group, since it was this attitude that had held the group together throughout the entire period of preparation.

Attitudes affect the behavior of an individual in relation to all eight subsystems of behavior. They serve a primary function of bringing together the diverse experiences to which an individual is exposed and forming them into a cohesive, organized whole. It is through the attitudes and belief systems of an individual that environmental perceptions acquire meaning. The danger is that attitudes may become so rigidly adhered to that instead of assisting an individual in understanding his environment and the events taking place within it, the attitudes themselves become the perception. Essentially, such a person has no contact with the environment in areas of intensely held attitudes, and is blind to what others see. Thus, it is impossible to rationally discuss the elements of the situation, because there is no common basis for a discussion.

Attitudes develop out of parental and group influences and from innate personality characteristics. Culture sets the limits for attitudes, but within these boundaries there is a wide variation allowed. The family is the initial limiting influence. Attitude toward food, elimination, expression of emotion, aggressiveness, hobbies, achievement, and so forth are developed as a child becomes aware of parental values. The parents may be extremely positive toward achievement behaviors and reward success within achievement areas, while they are negative toward wasting time. The child soon learns that if he is busy doing something, he will foster a positive attitude toward himself on the part of his parents, but that if he is sitting around doing nothing he will run into trouble. These expectations of the parents are subsequently modified by the influence of peer groups and by increased contact with people who hold a variety of different attitudes.

The process of changing attitudes is complex and time consuming. It requires that an individual objectively examine the critical elements of the attitude and identify those components that are valid and those that are prejudgments. Selected exposure to situations and persons involved in the attitude helps to discriminate valid from invalid areas. One of the premises of the movement to integrate schools is that basic attitudinal changes can be facilitated by enforced contact, rather than by attempts to modify attitudes in isolation.

SOCIOECONOMIC STATUS AND ROLE EXPECTATIONS

One way that people group themselves is in terms of their social class, a discrimination which is significantly related to economic factors. Within American society usually three classes are distinguished as representive of the entire population: upper, middle, and lower. Within each of these groupings there are further subdivisions into upper and lower subgroups. An individual may be placed in an upper lower-class designation, with the possibility of moving into the lower middle-class. In addition to economic factors, other criteria used in the designation of social class and status include race, occupation, type and location of residence, and educational level. People of the same class tend to have similar attitudes, values, and beliefs, to intermix socially and in business, and to marry into each other's families. Once in a while the Cinderella story comes true, but that is the exception, not the rule.

Many of the social movements within the United States have tried to eliminate the contradition between the democratic principle that all individuals are granted equal opportunity and the apparent restrictions within the environment that serve to prevent the realization of full equality. A major position of equal rights movements, be it a women's rights, black, Indian, Chicano, or gay liberation movement, is that every person regardless of his or her position in society, has the right and privilege to attempt to achieve any status or position that is available. For example, the women's liberation movement does not imply that achievement of their goals would require all women to accept full-time employment outside of the home, but rather, seeks the privilege for each woman of making this determination for herself, and the assurance that if she should elect to seek employment, she would receive equal consideration with all other applicants, regardless of race or sex. Any of these movements tend to stir up controversy by their nature of challenging the status quo and the existent social attitudes that underlie the problem.

Until recently in the United States, and still to some degree within various institutions of society, certain forms of achievement have been limited primarily to the white, Anglo-Saxon, Protestant male. Persons of certain nationalities and religions were excluded from even competing for selected goals, and if competition were allowed, it was under the conditions of a fixed quota. Irish Catholics were finally awarded full equality when John F. Kennedy became President of the United States, but Jews, Negroes, American Indians, and women are still barred from full and equal participation.

The roles an individual assumes at various times of the day also relate to his social class. Furthermore, the some role will have different expectations associated with it in different social classes. To be a housewife in the lower socioeconomic group would involve washing, ironing, cooking, cleaning, sewing, and other activities required by members of the family. The identical role in the middle class might differ because the housewife can purchase aids, such as frozen foods and electric brooms, which reduce the amount of time required to complete the tasks. For an upper-class woman, the role might refer to the fact that she has a housemaid and cook to whom she gives directions.

Social class and role expectations influence patterns of behavior indirectly. The higher the socioeconomic class, the greater the variety of choices of behaviors will be available to achieve specific goals. Any time the availability of resources become restricted, e.g., due to economical or geographical restrictions, the range of behavior will also become restricted. There is a close relationship between the strength of achievement behaviors and membership in a social class. It is for this reason that many have viewed higher education as a means for expanding the opportunities for social mobility. Education alone is not sufficient, however. An individual must develop behaviors that are appropriate to the social class to which he aspires, if he wants to achieve membership with a minimum of difficulty. This requires development of accepted behavioral patterns for all eight subsystems. One must learn to eat in the "correct" way, say the right things, know the right people, engage in the desired hobbies, and so forth. Essentially one must develop the behaviors of the class to which one aspires before becoming a member. The process of acquiring the different behaviors may also serve to isolate the individual from his original class values, as in the play *Pygmalion*, by George Bernard Shaw, the story of the transformation of a poor flower girl into a person of wealth and station. It is important to remember that what is transformed is the *behavior* of a person, not the person himself.

LOCUS OF CONTROL AND ALIENATION

An individual's perception of the environment is influenced by the degree of independence which that person experiences in terms of the dimension of control. Rotter proposed that behavior can be elicited by two types of cues: internal and external.[25] *Internal cues* refer to stimulus conditions that arise within the body and contain learned associative meanings, such as the relationship between tachycardia and anxiety. *External cues* refer to any aspect of an individual's environment, outside of the body, to which he may be responding at a given time and which has acquired meaning as a result of previous experience. An individual is experiencing both forms of cues constantly, and yet his behavioral response can be shown to relate most strongly to one of the two dimensions of cues. Rotter concluded that it was possible to differentiate two types of individuals on an external-internal locus-of-control measurement. The external locus-of-control individual is one whose behavior is primarily related to external environmental events, whereas in contrast the internal locus-of-control individual responds primarily to internal states of feelings and thoughts. In terms of behavioral system analysis, the manifest behavior of an individual who is externally controlled will demonstrate a closer relationship to the events transpiring in the environment than will the manifest behavior of a person characterized as having an internal locus of control.

Related concepts from the field of sociology are those of alienation and powerlessness.[26] *Powerlessness* is the expectancy or probability held by the individual that his own behavior cannot elicit the desired outcome, while the concept of *alienation* refers to the feelings of powerlessness, meaninglessness, nonnormality, value isolation, and self-estrangement. An individual who feels that he has no control over what happens to him also feels that other individuals in the environment have the control. Such an individual is high in terms of powerlessness and will behave in such a manner as to externalize his feeling of internal impotency. A hospitalized patient is often placed in a position of having little control over what may happen to him during the day. Those persons who are used to having control are likely to attempt to maintain it within the hospital, thereby establishing the potential for conflicts between patient and staff. A patient who is high on the dimension of powerlessness is unlikely to follow the teaching given by staff to provide him with the knowledge to prevent illness, since he feels that he has no impact upon his own function.

Locus of control and alienation are both related to the behavioral subsystems. Achievement behaviors are directed to fulfilling the goal of mastery over the environment and oneself. Individuals who perceive themselves to be mastered may develop one of two response tendencies. They may either attempt to gain, or regain, control of the situation, or they may remain in a position of submissiveness. Persons who are unable to control the environment may utilize a greater frequency of dependency behavioral actions. An external orientation requires greater attendance to environmental events, and a consequent greater expenditure of energy in terms of activity of the ingestive subsystem. It is important to remember that these are innate styles of perceiving the environment over which an individual may have little, direct conscious control. Nevertheless, they are important influences on the pattern of behavior in relation to the environment, and contribute to the overall perception and performance of the individual.

CREATIVITY AND PROBLEM SOLVING

People are confronted by problems and decisions throughout their lives. Society has developed a variety of solutions to problems commonly encountered by most members of the social group, but there are always some individuals who are able to develop unique and original approaches to these problems that may eventually be adopted by the society for use by all its members. These individuals serve the creative function of the group.

Creativity is, in part, the ability to find new solutions to old problems. It may be a creative explanation of why an event occurs, such as Newton's explanation of why an apple falls. It can be creative problem solving or the prediction of an event that has yet to happen as a result of knowing the conditions that currently exist. The overall goal of science is to be able to predict. A third type of creativity is the ability to invent a new set of conditions. It involves the least structured situation of the three requires the intuitive precognition of direction toward a goal, as in the example of the painter who begins with an empty canvas and a mental image of the finished painting.

It is necessary to distinguish between creativity in relation to the individual and creativity in relation to society. The latter is implied in the usual use of the term, whereby an individual is described as creative if he is able to invent, compose, or paint a creative product that is judged by the group to be so. Most individuals, however, are not creative in the sense that people will pay them money for the products of their creativity. However, the ability to express one's inner condition, be it through dance, installing plumbing, or meditating can be a creative experience *for the individual.* The end product may not be the same, but the process of creative expression can be just as rewarding and meaningful.

There is some experimental evidence that an important factor in creativity is the ability of an individual to tolerate ambiguity. Some people are able to cope with inconsistency, surprises, and uncertainty, while others find it extremely difficult to do so without disruption of usual behavior. A person who is able to cope with little difficulty most often is able to do so at the expense of creative problem solving, since the method by which the coping occurs is through the selection of the most likely solution; this solution will be repeated as long as it is functional. The behavior of such an individual will be relatively rigid, i.e., inflexible, and will be characterized by a frequent use of the same behavior. Other individuals will be intrigued by the inconsistencies and will attempt to arrive at a specific solution that is based upon the conditions. As they attempt to resolve the ambiguity of the situation their behavior may appear to be erratic and contradictory, but it usually follows a consistency that is within the individual. The inflexibility and rigidity of the person who is unable to tolerate ambiguity is a function of the aggressive-protective subsystem whereby the situation elicits the requirement for protection from perceived threat; in contrast the behavior of the second group of individuals is achievement oriented as they attempt to master the environment that poses a problem to be solved.

SELF-CONCEPT

The perception of oneself in relation to the environment is a powerful determinant of behavior. An individual's self-concept affects all subsystems of behavior directly and indirectly through its effect on what is perceived and on what is felt to be important to express through behavior. The term *self-concept* can be defined in a number of ways, but in general it is an abstract term that refers to a person's view of himself. It includes a person's physical, cognitive, and emotional characteristics, all of which are internalized. When you look in a mirror what do you see? Each person has an internalized body image that is related to, but not necessarily idential with, reality. There is no reason to assume that one's perception of oneself follows any other rules than those that govern the perception of others. Consequently, certain aspects are attended to while others are excluded, and various interpretations and meanings are placed on that which is perceived. Suppose much of your life you have had a weight problem, and suddenly instead of being overweight you are now underweight. There may be a time lapse between attainment of the new body profile and the corresponding change in body image. The experience of phantom-limb is another example in which the internalized body image may require an extended period of time to adjust to physical realities. In the example of the overweight/underweight person, there is a time during which the now underweight person still feels, and reports himself to be, overweight. Perception of the physical attributes of the self may result in labels such as beautiful or ugly, like or dislike, and may affect the behavior of the individual, particularly in interpersonal situations.

The psychological aspect of the self refers to the way in which an individual perceives himself and his habits, behavior, abilities, skills, knowledge, and all other internalized capabilities. Whether an individual will share these self-perceptions with another depends on a great many factors, not the least of which is the nature of the relationship between the individuals. That aspect of ourselves that we are willing to share with others is the *persona*, or the part we play with others. The persona is significantly determined by the situation. For example, a person will expose different aspects of himself at a social occasion such as a wedding or party than at the office of his professor. A woman may allow her husband to become familiar with characteristics of herself that she would not share with others, even with her child. The impressions conveyed to others of what we are as people are just as real in both situations, but the nature of the situation and the relationship between the individuals will govern how much and what we are willing to expose to the outside. A patient in a hospital frequently behaves in the way in which he thinks he is expected to, rather than as he desires. The behavior seen by the nurse consequently relates to the persona of the individual, which may be either distant from, or close to, the individual's self-concept.

The rationale underlying the sharing of oneself with another person is a major determinant of behavior. An individual will share his thoughts, feelings, and life experiences in much greater depth with a psychotherapist, or even with a friend, than with a casual acquaintance because the friend or therapist is seen as someone who can be of assistance in some manner. If the assistance is related to gaining increased self-knowledge, it is imperative that the sharing of self-concepts occur on a relatively deep level.

The purpose for increased self-knowledge may be to create a more adaptive relationship with oneself or others in the environment. Whether an individual's behavior will change depends on how maladaptive the behavior is in relation to his self-concept. A major criticism by family and friends of a person in psychotherapy is that they perceive no difference, or a change for the worse, in that person's behavior. In such situations it must be remembered that the relationship between the manifest behavior of an individual and the internal perception of the self may have been incongruent before therapy, even though the manifest behavior was socially desired and accepted as healthy. Suppose an individual begins to express a great deal of angry behavior toward relatives after having been in psychotherapy for a time. The recipient of the anger may judge the therapy to have been harmful and useless to the patient, whereas the individual may be expressing in his behavior the true state of affairs, which he had failed previously to act upon.

SUMMARY.   This chapter has focused on general biological, psychological, social, and cultural regulators that function to limit and influence activity within the behavioral system. The regulators discussed include: genetic inheritance; circadian rhythms; age; sex; attitudes, values, and beliefs; social class and role expectations; locus of control; creativity and problem solving; and self-concept.

These regulating factors are relatively stable, enduring traits or characteristics of the individual that are manifested in an extensive variety of dissimilar situations. Some of these characteristics develop from an interaction with the environment, e.g., social class and role expectations, circadian rhythms and attitudes, but are eventually internalized within the individual's personal system of characteristics and become virtually autonomous and independent of the environment. Others, such as sex and genetic inheritance, are givens at the time of birth. All these regulators share the ability to influence the goals, set, choice, and acts associated with each of the behavioral subsystems as well as the information that is made available to the system as a whole. It is for this reason that the characteristics are treated as a class of intervening variables, because their activity influences both the input and output of the behavioral system.

STUDY QUESTIONS

1. What is the nature of the interaction between an individual's attitudes, beliefs, and values, and the situations that are actively sought? actively avoided?

2. How does your behavior change in relationship to the person with whom you are interacting? the situation?

3. Examine the experience of hospitalization in terms of those factors that assist an individual to maintain control within the situation.

4. What conditions existing within a hospital may create a feeling of powerlessness in a patient?

seven | *Behavioral Assessment and Nursing*

Throughout his lifetime, an individual will encounter and interact with an infinite variety of persons. Many interactions are limited to one time and are defined by a specific purpose on the part of the participants. Others, in contrast, may last throughout a lifetime, such as relationships with family and close friends. Interactions with people who are known and familiar are characterized by an initial exploration of how the person appears to be feeling that day. How is their behavior today as compared to yesterday? the week before? ten years ago? In other words, the focus is on how the observed behaviors at the time of interaction compare to past behaviors. Implicit in this process of comparison is the knowledge of what is "usual" or "adaptive" behavior for this individual, based upon the extended relationship.

When one encounters a person for the first time, this frame of reference of "usual" behavior is absent. The only information available is how that person is behaving when the initial contact is made. If the conditions surrounding the interaction are formalized and socially defined, they provide a frame of reference for assessing the adaptiveness or maladaptiveness of the observed behavior. For example, if you are shopping in a department store and interact with a salesclerk, the situation of shopping and the formal expectations of role behavior for salesman and buyer establish a framework for assessing the behavior contained within the interaction. One expectation is that the salesperson is oriented toward promotion of sales. Should the salesperson not promote sales, but rather discourage the buyer from purchasing a product, a question may be raised as to the underlying motivations of this apparent maladaptive behavior. Behavioral assessments of initial and one-time interactions are limited to the apparent appropriateness of the observed behavior of all parties in terms of the situation in which the contact takes place.

The nurse-patient relationship can be viewed as a composite of initial and extended interaction. Initial interactions usually are based on existing behavioral patterns, and future interactions expand on this foundation. The nurse, however, is concerned not only with the presently observed behavior but also with the characteristic patterns of behavior that existed before the initial contact with the patient. In this sense, the initial assessment is concerned with the acquisition of knowledge about the patient and his behavior, in order to artificially produce an extended relationship between nurse and patient. The nurse must assess not only the behavior but also the situational characteristics that tend to favor the use of one behavioral pattern over another. In this way, it becomes possible to evaluate changes in behavioral actions, and their adaptiveness or maladaptiveness, based on knowledge of pre-illness patterns of behavior.

The focus of the formal assessment process is to obtain knowledge regarding a patient through interviews and observations of the patient and his family. The purpose is to evaluate the present behavior in terms of past patterns, to determine the impact of the present illness and/or hospitalization on behavioral patterns, and to establish the maximum possible level of health toward which an individual can strive. The assessment is both predictive and descriptive in nature; it describes the point on the health-illness continuum at which the patient is currently located, and predicts the degree of expected movement in the direction of health, or wellness. A prediction of the degree of improvement must consider any potential medical complications that might occur. Total independent function may not be a realistic goal for all individuals; however, adaptive patterns of behavior that account for the restrictions in choices available to the individual should be a realistic outcome for anyone, regardgardless of pathology.

In Chapter 1 wellness is described as an integrated state of optimal function that includes the psychological, social, and cultural dimensions contained within the individual. It is also proposed that these four dimensions are made apparent through behavior, and that changes within any or all the dimensions would result in a change in behavior. The assessment of behavior must include all these dimensions in order to establish those areas in which medical and/or nursing assistance is required to attain maximum wellness. This requires consideration of the entire individual, including the biological, psychological, and sociocultural characteristics of the patient and their relationship to behavior; the impact of illness upon behavior; the impact of the environment in which the patient is located, e.g., home, hospital, nursing-care home; and the potential impact of the environment to which the individual is expected to return. What may be healthy behavior within the hospital may be exceedingly maladaptive in the normal environmental conditions in which the individual lives and to which he plans to return, following hospitalization.

The behavioral systems analysis approach provides a comprehensive framework in which these various types of data can be organized into a cohesive structure. Observations of current behavior can be related to information obtained from a formal interview protocol that examines the environmental influences, regulatory mechanisms, and personal structure and function of the eight subsystems of the individual. Knowledge of the relationship between current behavior and past behavior allows the identification of problem areas in which the patient may require assistance from the nurse and other health professionals in order to modify or acquire adaptive and functional behavioral patterns. Furthermore, this knowledge provides a frame of reference for establishing realistic goals for recovery based upon an appraisal of the patient behaviors prior to the onset of the disease.

A nursing assessment tool must be designed in such a manner as to be both reliable and valid. *Reliability* refers to the accuracy of the information obtained through the use of the tool, in the sense of stability and reproducibility. The basic question is whether the tool, if administered twice under identical conditions, will result in identical responses. Failure to demonstrate reliability may be due to the patient population on which the tool was tested, or because of poor item construction. For example, suppose that five different interviewers each asked a person how many meals he usually eats in one day, and his responses were three, one, three, five, and ten. The obvious inconsistencies in the response elicited by the question is an indication of apparent unreliability. The item may be unreliable because the patient has deliberately distorted his responses, or he may have misunderstood the question; it is also possible that the individuals who asked the question did so in an inconsistent manner. If the reason for the inconsistency is that the respondent misunderstood or misinterpreted the question, then the problem of unreliability may be resolved by rewording the item to prevent confusion. Should the reliability be a matter of the way in which the interviewer elicited the information, then additional training and experience in the practice of administering interviews would be warranted. It is possible to test the reliability of an item by using an alternate which is designed to elicit the identical information from the patient, but which is worded differently. For example, an item worded "How many meals do you eat each day?" would be expected to elicit the same information as an item that asks, "Many people eat three meals each day. How many do you eat?"

An additional source of data that can be used to establish the reliability of information obtained through an assessment questionnaire is to refer to the information obtained from members of other disciplines—medicine, psychology, social work, for example. A patient may be interviewed by a number of individuals, each of whom elicits information specific to his discipline, as well as a general data base that is interdisciplinary in nature. The information obtained from the patient should demonstrate a high degree of consistency over many interdisciplinary interviews.

Validity is a measurement of the extent to which an instrument or tool does what it states its intentions to be. If the tool is intended to measure behavior it must demonstrate that it does measure behavior, as purported. The simplest form of validity is based on a superficial examination of the content of the items. This type of validity is called *face validity* because the items appear "on the face of it" to be valid. For example, items that relate to diet and food preferences would appear to be related to ingestive behaviors. *Content validity* involves the use of empirical data to support the items that are included in the test. For example, it has been reported that individuals who cope with anticipated stress through the use of denial are less able to respond in an effective manner to the stress when it occurs. Based upon this report in the literature, an item might be included to assess an individual's coping mechanisms in those situations when stress of some form is anticipated. The sample assessment tool discussed later in this chapter has content validity because the items are based upon an examination of the literature on various areas of behavioral content. The most sophisticated demonstration of validity is the ability of the instrument to *predict* subsequent behavior, which is an ultimate goal of nursing assessment tools.

The specific format used for construction of the assessment tool will affect the depth of information elicited through the interview. Observations are best suited for behavioral description and evaluation, but they are limited in the depth of specific information that can be acquired. Structured interviews and questionnaires can be developed that will elicit very specific information. Marshall and Feeney evaluated the effectiveness and efficiency of structured versus unstructured interviews.[1] Their results indicated that the structured method yielded significantly more information and required less time to administer than did the unstructured, intuitive approach. The disadvantage of a highly structured assessment form is that the data obtained may be limited to those that are specifically requested by the items that are included. The expert clinician may utilize the structured items for inquiry and should be able to elicit additional information in those areas in which there appear to be maladaptive or malfunctional behavioral patterns. A combined structured/unstructured format may be the most effective approach.

Once the information has been obtained it is necessary to formulate the various bits of data into a comprehensive statement, or *nursing diagnosis*. The function of such statements is to facilitate the development of a nursing care plan that is meaningful to the patient as well as to all personnel who provide care for that individual. Those areas in which the individual is experiencing maladaptive behaviors should be thoroughly evaluated for the probable source of difficulty. This requires examination of the general regulators; the specific regulatory mechanisms for the subsystem involved in behavioral disturbance; the structural components of the subsystem; the situational and stress components that might be creating behavioral disturbance; and the relationship between activity in the focal subsystem of disturbance and activity in the remaining seven subsystems.

For example, obesity as a symptom of maladaptive ingestive behaviors may be due to:

1. insufficient economic resources to buy foods that contain protein, i.e., the general regulator of socioeconomic status;
2. decreased metabolic requirements due to inactivity, i.e., the specific regulator of ingestive activity;
3. continuous eating without awareness of the amounts of food consumed in 24 hours, i.e., perseveratory set;
4. food preferences related to high-calorie foods, i.e., choices;

5. lack of encouragement by persons in the environment to reduce consumption of foodstuffs in order to lose weight, i.e., sustenal imperatives;
6. compensation for difficulties in relating to others, i.e., a disturbance focused in affiliative behaviors that is compensated for by increased activity of the ingestive subsystem.

It is apparent that any one, or combination, of these disturbances is sufficient to produce a state of obesity. A general intervention may be effective for any number of reasons. However, it may be more effective and efficient to identify the focus of disturbance and design interventions that are specific to the disturbance. If the interventions are established without a thorough evaluation of the specific origin, or origins, of the problem, the interventions may be unsuccessful.

It is also important that attention be directed to the identification of patient strengths and existing adaptive behaviors. All too often the problems of a person are attended to, but little consideration is given to the basic areas in which adaptive behaviors are present. These behaviors are as important to an individual as are the maladaptive behaviors, and the nurse must make provisions for assisting the patient to *maintain* these healthy behaviors during recovery. It is possible that interventions that strengthen healthy behaviors may be more effective in the recovery process than those that deal directly with the maladaptive behaviors.

The situational factors of the hospital setting must also be considered when one assesses a patient's behavior. Behaviors that may be adaptive and functional for an individual during his normal, daily life may not be adaptive to the conditions of the hospital. The normal hours of sleep—10:00 P.M. to 7:00 A.M.—may not be the usual hours of sleep for a nightwatchman. When that individual is hospitalized, it is necessary for him to adjust his adaptive pattern of restorative activity to one compatible with the activity of the hospital setting. In such a situation it is necessary for the nurse to assist the individual in making temporary adjustments in his normal pattern.

The question that must be considered is one concerning the amount of information desired from the assessment process and its relationship to the amount of time available for this process within a particular clinical setting. It is fair to say that if something is felt to be important, time will be made available for its implementation. A formal assessment protocol may require from 15 to 60 minutes for the initial gathering of information, depending upon the amount of information required from the individual, the complexity of problems identified through the process, the expected length of stay, the expectation of extended contacts in the future, and the depth of detailed information desired by the nursing team. The time devoted to this interview can save inestimable hours of planning and can avoid the implementation of insufficient, or inappropriate, interventions. Interventions that are based upon a thorough, comprehensible assessment of the patient, in which attention has been paid to the aspects that influence the entire person in the hospital, as well as in the personal setting, are much more likely to have a long-range impact. It is important to remember that this initial process also assists in the establishment of a nurse-patient relationship in which both parties begin to relate to one another, thereby assisting the patient in establishing an affiliative relationship with a member of the health team. The assessment process will be effective only if the information that is acquired is seen as both valuable and useful by all participants. If the information is not utilized in the formulation of an individualized nursing care plan then the purpose of the process must be closely examined.

AN ASSESSMENT TOOL BASED UPON THE BEHAVIORAL SYSTEM APPROACH

The assessment protocol which follows will not be strikingly different from the usual assessment questions. There is just so much information that is relevant to the nursing process. The primary difference between this tool and others available in the literature is its relationship to the behavioral systems approach. The main purpose of each assessment item is the acquisition of specific knowledge regarding the structure and function of the eight subsystems of behavior, and those general and specific factors that influence and modify behavior for a given individual. This protocol has intentionally been constructed for use in any area of nursing, but it could be modified in terms of the specific clinical setting, the individual patient and his medical diagnosis, and the specific information desired by the nursing team.

The assessment tool has been divided into two major sections. The first section is concerned with the acquisition of data related to the general regulators of behavior. This includes analysis of the physical environmental factors, perceptual process regulators, and general characteristics of the individual. The specific purpose of each question is indicated in parentheses, following each question. The second section is focused on the analysis of the function and adaptiveness of the eight subsystems. Mischel suggests that the only way in which it is feasible to determine the stimulus conditions to which an individual is responding is by actively enrolling him as a collaborator in the assessment process.[2] Together with the observer, the individual can provide possible hypotheses regarding the conditions that result in the presence or absence of various behaviors. This is the intent of the following tool: to reveal through the collaborative efforts of patient and nurse the meaning and function of behavior.

I. Items related to general regulators of behavior

1. What is your name? (Self-concept)
2. What is your birthdate? (Age)
3. Where were you born? (Sociocultural environment and regulators)
4. Are you married? (Role expectations; specific regulator of sexual, affiliative, dependency, and achievement subsystems)
5. If not currently married, have you been married in the past? (See question 4)
6. Do you have children? If so, how many? If not, what are your intentions regarding establishing a family in the future? (Sociocultural values; role expectations; specific regulator of sexual subsystem)
7. Who are the members of your immediate household? (Socioeconomic status; sociocultural environment; specific regulator of affiliative and dependency subsystem)
8. Describe your home and the neighborhood in which you reside. (Physical and sociocultural environment; socioeconomic status)
9. What is your highest level of academic preparation—e.g., highest grade or degree obtained? (Genetic inheritance; socioeconomic status; creativity and problem solving; specific regulator of achievement subsystem)
10. What is your current occupation? Where are you employed? How long have you been engaged in this occupation? Have you ever been active in a different occupational field? If yes, why did you change? If no, have you ever considered changing your career choice? (Socioeconomic status; creativity and problem solving; self-

concept; specific regulator of achievement subsystem; also specific regulator of sexual subsystem of females)

11. What church or religious denomination do you belong to as a member? Are you active? Are there specific beliefs that you adhere to? If you do not subscribe to a particular religious creed, would you describe your basic beliefs which provide your ethico-moral structure? (Sociocultural environment; attitudes, values, and beliefs; specific regulator of ingestive and affiliative subsystems)

12. What season of the year, if any, would you describe as the most comfortable for you? During which one are you least comfortable? (Physical environment; circadian rhythms)

13. Would you describe yourself as being extremely sensitive to light and noise under normal circumstances? At this time? If more sensitive at this time, do you feel that it is related to your illness or to conditions existing within the hospital? (Perceptual sensitivity; physical and sociocultural environment; interpretation and labeling of perception; general regulator of set; specific regulator of ingestive subsystem)

14. Do you find it difficult to concentrate in a noisy room? (Perceptual sensitivity; ability to exclude surrounding stimuli)

15. In general, how do you respond to situations that require you to do something which you are reluctant to do? (A) Do you plunge in and complete it as soon as possible? or (B) Do you put it off as long as possible? (Perceptual coping style—response A would indicate an augmentor or vigilant coping style, whereas response B would indicate a pattern of denial or defensiveness)

16. What situations make you feel anxious? (Perceptual coping style; interpretation of environmental conditions)

17. What situations make you feel calm and secure? (Perceptual coping style; interpretation of environmental conditions)

18. Have you ever been hospitalized before? What do you expect to have happen to you during this hospitalization? Is there anything that has happened to you thus far which has been confusing or for which you would like an explanation? What does it mean to you to be in the

hospital? (Perceptual labeling and interpretation; coping style; attitudes, values, and beliefs; locus of control; expectancies)

19. Would you describe your physical appearance as being similar to your father's or mother's? Are there any problems of genetic trait inheritance in your family? (Genetic inheritance; self-concept)

20. What are, or were, your father's and your mother's occupations? (Genetic inheritance; socioeconomic status; sociocultural environment; specific regulator of achievement subsystem)

21. How would you describe your attitude toward life? What are the most important values to you? (Attitudes, values, and beliefs; role expectations; self-concept)

22. Do you feel that you have control over what happens to you? What does the term *fate* mean to you? Has this hospitalization affected your feelings of control or of lack of control? (Locus of control and alienation; self-concept; attitudes, values, and beliefs)

23. How would you describe yourself? What do you like best about yourself? If it were possible, what is the primary aspect of yourself that you would like to change? Do you share your feelings with another person with ease, or are you highly selective as to whom you will share with? (Self-concept; specific regulator of affiliative subsystem)

24. Do you feel most alert in the morning or in the evening hours? (Circadian rhythms)

25. What do you think is meant by the term *good patient*? Is it important to you to be like that? (Role expectation; perceptual meaning; attitudes, values, and beliefs; self-concept)

26. Has this illness affected your life in any way? What does this illness mean to you? What do you hope to gain from this illness? Do you think this illness will affect your future? If so, in what way? (Attitudes, values, and beliefs; expectations of situation; problem solving; self-concept; locus of control and alienation)

27. Observations related to general regulators of behavior: Note the sex and ethnic identity of patient. It not immediately apparent by observation, ask a direct question.

(Genetic characteristics; sociocultural environment; attitudes, values, and beliefs; self-concept; specific regulators of all eight subsystems)

Does the appearance of the individual confirm his reported birth date? (Age; sociocultural regulators)

What mannerisms accompany the responses? Do particular questions elicit behavioral manifestations of discomfort such as flushing, hand movements, sweatiness of palms, or lack of eye contact? (Identification of potential areas of disturbance)

Does the patient use the word "they" instead of "I" when responding to questions? (Locus of control and alienation)

II. Items related to specific subsystems of behavior

A. Ingestive subsystem

1. Regarding ingestion of food material:

   How many meals do you eat each day?

   What times do you prefer to eat?

   What foods you prefer for breakfast? lunch? dinner?

   What food do you particularly dislike?

   Do you like to snack between meals? before bed?

   Do you tend to finish all the food before you, or do you eat until you are full and then no more?

   When you are eating, do you prefer to eat by yourself? with others? in front of the television? while reading a book?

   Do you have any physical limitations related to foods, such as allergies, false teeth, digestive problems, nutritional deficiencies?

   Has this illness affected your usual pattern of food intake? If so, in what way has it been altered? Are there foods that you ordinarily do not eat which you have desired at this time? Foods that make you worse? Changes in appetite?

   In general, what significance does food have for you?

2. Regarding ingestion of fluids:

   What types of fluids do you prefer?

   Do you drink fluids such as coffee, soft drinks, milk, juice between meals? If so, how often and what specific forms?

   Do you drink alcoholic beverages? If yes, under what

conditions? (If response indicates that alcoholic intake is related to meals then' the behavior is related to the *ingestive* subsystem; if restricted to parties and social occasions, then the behavior is *affiliative* in nature; if alcohol is used to relax following periods of tension, then the behavior is *restorative*.)

Since you have become ill have you noticed any change in your preferences for fluids? the amounts you have been drinking?

3. Regarding sensory modalities:
   Do you have any sensory limitations or restrictions related to:
   a. Vision?
   b. Hearing?
   c. Taste and smell?
   d. Touch?
   e. Pain?
   f. Perception of temperature?
   Do you have difficulty tolerating noise? quiet?

4. How do you seek information about areas in which you are unfamiliar? by asking question? reading books? taking classes?

5. Observations related to ingestive behaviors:
   General nutritional status: appearance; skin color; condition of teeth.
   Respiration: depth and quality of inspiratory phase.
   Ability to comprehend intent of questions.
   Nonverbal behavior indicating potential sensory deficits.
   Level of consciousness: orientation to time, place, and setting; alertness or confusion; speed of response to verbal stimuli; responsiveness to physical stimuli, e.g., a neurological exam.

B. *Eliminative subsystem*
   1. Regarding bowel function:
      How often do you usually have a bowel movement? What time of day?
      Are there any foods or fluids that help to stimulate a bowel movement? Do you use laxatives, stool softeners, or enemas to initiate a movement? If yes, how often? Have there been any changes in your bowel movements

related to this illness?   If so, describe, i.e., frequency, color, consistency.

2. Regarding urination:
    How often do you urinate?
    Have you experienced any difficulty in retaining urine? voiding?
    Has this illness affected your pattern of voiding in any way, e.g., in frequency, amount, or color?

3. Regarding menstruation:
    How often do you menstruate?
    What is the length of your monthly period?   Are you regular?
    Do you experience any discomfort in relationship to menses?
    Have you observed any change in your menses as related to this illness?   If so, please describe.

4. Regarding expression of affect:
    What do you usually do when you are upset and angry?
    What do you do when you feel happy?
    Do you readily share your feelings with others?
    Is it easier for you to express yourself through words? pictures? writing? music?
    What happens if you try to hold your feelings to yourself?
    Have you experienced any changes in your feeling responses to situations as a result of this illness? If so, please describe.
    What situations are easiest for you to express your feelings in?
    Do you ever cry?
    Are there certain individuals with whom you are more comfortable expressing feelings?

5. Observations related to eliminative subsystem:
    Respiration: depth and quality of expiratory phase.
    Nonverbal behaviors and degree of congruency to stated verbal content.
    Appropriateness of stated affect and verbal content.
    Comfort and ease of speech.
    Quality of responses to questions; types of words used.
    Any special areas of elimination related to medical problem, such as feces, urine, wound drainage, vomitus, bleeding, body temperature, sputum, and menstrual flow.

C. *Sexual subsystem*

1. What does the word *masculine* mean to you? What does the word *feminine* mean to you? How masculine or feminine would you describe yourself?
Is being a mother/father important to you?
Describe your relationship with your wife/husband.
Describe your relationship with your children.

2. What is your normal frequency of intercourse? Masturbation? Has this illness affected your sexual activity? If so, in what way?

3. Are you comfortable in relating to persons of the opposite sex? same sex?

4. Are you currently, or have you ever been pregnant?
What does menstruation and pregnancy mean to you?

5. Observations related to sexual subsystems:
Presence or absence of secondary sexual characteristics.
Type of clothing worn by patient.
Use of make-up, hair style, shaving, and other sex-related activities.
Mode of interaction with persons of the opposite sex, i.e., touching, kissing, holding hands, teasing, seductiveness, verbal suggestiveness.
Affective response to pregnancy, or failure to become pregnant.
Extent to which behavior conforms with sex-role stereotypes.

D. *Affiliative subsystem*

1. Who is the most important person (significant other) in your life? Is this person a family member? a friend? What are the qualities of this person that you particularly enjoy in your relationship?

2. How would you describe the importance of your family life at this time? How often do you meet with your close relatives?

3. To how many organizations do you belong? Are you actively participating in these organizations?

4. What is the nature of your relationship with your neighbors?

5. Would you describe youself as a person who enjoys in-

teractions with others on a frequent basis? once in a while? never?

What do you do when you are by yourself?

Do you like to attend social functions? by yourself? with others?

Which do you prefer as a form of entertaining—a small dinner party or a large gathering?

6. How many intimate friends would you estimate you have? What are the characteristics of your relationship with these individuals?

7. Tell me about your relationship with your siblings?

8. Do you expect visitors while you are here in the hospital? family members? friends? priest, rabbi, or minister? neighbors? representatives of organizations of which you are a member?

Are there any persons whom you would prefer not to have visit during your hospitalization?

9. Has this illness affected the quality or quantity of your relationships with others? If so, in what way?

10. Observations related to the affiliative subsystem:

Use of supplemental, or substitute, forms of affiliation; via books, TV, roommate.

Quality of interaction with visitors.

Comfort and ease demonstrated in initiating conversations with others.

E. *Dependency subsystem*

1. Would you describe yourself as being independent or dependent upon others? Are there any activities with which you require assistance?

2. Do you usually see a doctor when you are feeling unwell? When you are experiencing pain do you prefer to be alone or with others?

3. How self-reliant are you? Is it difficult for you to request assistance?

4. Have you ever considered suicide? Under what conditions would you feel that suicide was the only solution? Have you ever actually attempted suicide? If yes, how and when?

Are you currently considering suicide? If yes, do you have a plan? What is it?

5. Has your present illness required you to be more dependent than usual upon others? If yes, whose assistance have you sought most frequently?
    Do you anticipate that you will require assistance after you return home?

6. Observations related to dependency subsystem:
    Frequency with which patient uses call bell for non-emergency requests.
    Frequency with which the patient seeks reassurance, verbally or nonverbally.
    Areas of activity in which assistance is required: eating, elimination, bathing, dressing, grooming, ambulation, administration of medications, selection of diet, etc.
    Ability to be independent in above areas of activity but refusal to assume independent action.
    Selection of certain objects for security, e.g., smoking cigarettes.

F. *Aggressive-Protective subsystem*
    1. When you feel threatened, how do you tend to respond to these feelings? What do you usually do when faced with a difficult situation?

    2. How soon after you began feeling ill did you make an appointment to see your doctor?
        Do you have a yearly physical examination?
        How often do you see your dentist?
        Have you maintained your immunizations?

    3. What do you usually do when you feel rejected by others? Do you experience these feelings very often?

    4. Has this illness made you feel threatened in any manner? How do you feel about the proposed treatments? surgery? medicine?

    5. If someone shouts at you, how do you respond?

    6. Would you like more detailed information regarding your care and proposed treatment?

    7. Observations related to aggressive-protective behaviors:
        Quality of responses to above questions in terms of denial; projection; rationalization.
        Withdrawal from persons in the hospital.

Degree of muscular tension—i.e., does patient appear to be poised for flight or fight?

Presence of startle responses when approached.

G. *Achievement subsystem*
1. What do you hope to achieve in your lifetime?
   How important is this goal to you?
   At this time, do you feel that you will be able to accomplish this within your lifetime?
2. Do you prefer to be in control of situations? Do you enjoy competitive situations? Are you a good loser?
3. Are you mechanically inclined?
   Do you prefer activities that require intellectual skills? Or do you prefer activities that are artistic and creative in nature?
4. What motivated you to select your occupation? If you were able to do anything you desired, what might that be? How satisfied are you with your opportunities for leadership and advancement?
   Do you feel that your performance at work is valued and noticed?
5. Do you consider yourself to be a lucky person? Tell me in what ways you have been lucky in the past.
   Do you like to attempt difficult, challenging tasks that you may not be able to master?
6. Has this illness affected your ability to work or to engage in those activities that are meaningful to you? If yes, in what ways have you been affected?
   What are your goals for this hospitalization?
   What are your goals for recovery from this illness?
7. Observation related to achievement behaviors:
   Concern with maintenance of control of hospital situation.
   Degree of compliance or noncompliance with medical and nursing treatment program.
   Ability to formulate realistic goals for hospitalization and recovery.
   Relationship between expressed interests and occupational goals.

H. *Restorative subsystem*
1. At what time do you usually retire? arise? How many hours of sleep do you need? Do you sleep throughout

the entire night?   If no, what events tend to waken you?
Do you have difficulty going to sleep?   If yes, what assists
you in falling asleep?
Do you dream at night?   Tell me something about their
quality.
What activities do you engage in before going to bed—e.g.,
bath, reading, drinking hot liquids, sleeping medication?

2. Do you rest during the day?   If yes, how often and for
   what length of time?

3. Have you experienced any difficulties related to sleep and
   rest since you have become ill?   If yes, tell me about these
   problems.

4. Tell me how you spend your leisure time, for example the
   physical activities in which you participate.
   Do you prefer sedentary activities or physical activities?
   How much time do you spend each day in leisure activities?
   Do you feel better, i.e., more relaxed, following these rest
   periods.

5. When you are feeling ill, are there any particular activities
   that make you feel better?

6. How relaxed would you describe yourself?

7. When was your last vacation? Tell me something about
   the vacation and how you spent your time.

8. Observations of restorative behaviors:
   Observe sleep patterns.
   Expressions of fatigue; lack of energy.
   Active involvement in treatment plan and expression of
   concern for recovery.

ALTERNATIVE APPROACH TO THE
ASSESSMENT PROCESS

Once a nurse has become proficient in interviewing and in knowing
what specific areas of behavior to review during the assessment, a
more open-ended approach to the wording of questions may be
preferable. Open-ended questions that begin with unstructured
stems, such as "Tell me about . . ."or "In what ways has this illness
. . ." permit an individual to interpret the intent of the question on
a personal level and provide a wider range of possible responses. An
interviewer must be able to note the specific areas that are not men-

tioned by the patient, in order to include more structured questions and cover the entire range of behavior. Open-ended questions are most effective with a low-anxiety, verbally proficient adult. Patients who are highly anxious or who are limited in their ability to communicate, e.g., who speak a foreign language, or children, may be more comfortable with a highly structured interview protocol that elicits yes-no answers such as were outlined in the preceding section.

ANALYSIS OF ASSESSMENT INFORMATION

Once an interview has been completed and the initial observational data have been obtained, it is necessary to organize the information in such a way that the characteristic patterns of behavior for the individual can be determined, and the potential impact of illness and hospitalization can be examined. The purpose of the analysis is to identify those patterns of behavior that are adaptive and functional for the individual and that can be maintained within the constraints imposed by the illness and/or hospital setting; those patterns of behavior that are adaptive and functional for the individual but that must be temporarily modified or altered in some form to be adaptive to the limitations imposed by the illness and/or hospital setting; and those patterns of behavior that are maladaptive and/or malfunctional for the individual and that require an active plan of nursing intervention.

Adaptive behavior is assessed by the environment in terms of meaningfulness in relationship to the constraints of the external situation. For example, a pre-illness pattern of eating behavior may be to eat a balanced diet in three meals a day. Should the individual attend a brunch on one day, the total number of meals may be two, while the number of calories and dietary constituents may be constant. The individual has adjusted his ingestive behavior to the unusual external situation of a different meal pattern, while maintaining constant intake. If the individual attended the brunch, and elected to eat a third meal, and consequently a greater quantity of food on that day, the behavior would not be adaptive to the constraints of the environment, although it may be functional to the individual if his goal is to gain weight. Maladaptive behavior is characterized by an absence of immediate meaning and purpose, or by an intensity of response that is excessive or deficient to the demands of the environment. For example, the individual described above may suddenly stop eating at any time of the day, even when highly preferred foods are presented to him; simultaneously he may engage in a strenuous program of physical activity which should have the effect of increasing the amount of food normally ingested. Such behavior would be classified as a maladaptive pattern of ingestive behavior.

The function of a behavior is judged by whether that behavior is congruent with the stated goals of the individual. The analysis of functionality involves the inspection of the structural components of each subsystem in order to determine the degree of congruency among goal, set, choice, and action. If there is any discrepancy, the behavior will be malfunctional for the individual. For example, the goal of the behavior may be discrepant from the actions, such as when an individual eats in order to fulfill the goal of affiliating himself with persons, objects, and ideas. The discrepancy may be between the goal and set; goal and choice; goal and action; set and choice; set and action; and choice and action. Furthermore, it is necessary to establish whether the discrepancy is restricted to a single subsystem, or whether the problem involves an interaction between two or more subsystems of behavior.

When one analyzes discrepancies that involve more than one subsystem, the concepts of *compensation* and *insufficiency* may be applicable. *Compensation* refers to the process by which the structural components of one subsystem may be employed by an individual to attain a desired outcome that is not directly related to that subsystem. For example, an individual who has an insufficient number of choices available to meet the goal of the dependency subsystem, e.g., the lack of assistance by others, may compensate through the use of choices that are functional to the goal of achievement, e.g., mastery of the environment. *Insufficiency* is complementary to the concept of compensation, and refers to a subsystem, or component of a subsystem, that is underdeveloped or restricted in activity for any reason.

A detailed analysis of the assessment interview data and observations should determine the level of adaptiveness and functionality of the behavioral patterns of an individual in relation to the hospitalization, as well as before the onset of illness. Attention should be paid to those behaviors that are reported to have changed, as well as to those which have been stable, because each is important to the overall function of the behavioral system. It is possible, for example, that a recent change in behavior due to significant life events was a precursor of the illness.[3] Those behaviors that have not demonstrated any recent change may reflect a failure to adapt to new situational constraints and may require modification to promote health. The summary of the analysis should state the nursing diagnosis of current behavioral system function and of potential direction of change.

Preliminary Analysis of Regulatory Factors

A regulator, by definition, regulates by limiting the input received by the behavioral system. The basic question that must be answered in an analysis of regulatory factors is *how* the regulator limits and directs the observed behavior. How does the age and developmental status of the individual limit and direct his behavior? Children are not capable of engaging in the same actions as an adult and this limitation is a consequence of physiological, psychological, and sociocultural regulators. What physical or psychological deficits are present that limit the range of possible behaviors available to a given individual? What effect does the individual's family type, religion, economic status, and social status have on behavior? Theoretically, all biological, psychological, social, cultural, and environmental regulators could be examined for their impact upon the behavior of an individual. Such an analysis would be impractical; it would require an extensive knowledge of a wide variety of subject areas and would take too much time. The analysis, however, should cover the most important regulators of observed behavior. Some suggestions follow.

physiological: What is the impact of *age, sex, ethnic origin, physical limitations, internal physiological processes,* and the *level of developmental maturation* on the behavior of the individual?

social: What effect do *family structure, occupation, economic factors, residence, geographic location, social class, political beliefs,* and *other social factors* have on the behavior of the individual?

cultural: What effect do an individual's *value system, language, creative expression, custom, attitudes,* and *religion* have on the behavior of the individual?

psychological: What effect do *emotional states, perceptual styles, coping styles, cognitive and intellectual development, past learning history, self-concept,* and *personality* have on the behavior of the individual?

environmental: What effect do the *weather, altitude, physical location, time of day, temperature,* and *structure* of the environment have on the behavior of the individual?

general: What is the *nature of the interaction between* the *physiological, psychological, social, cultural,* and *environmental influences?* How does this interaction serve to limit and direct the behavioral action of a given individual?

Preliminary Analysis of Structure and Function of Subsystems

The second level of analysis deals with the individual subsystems of behavior, and determines whether there is congruency among all the structural components in their functional interrelationships. Congruency will be evidenced by stable, patterned behavior, whereas discrepancy among the various components will be evidenced by unstable and disorganized behavior. In analyzing subsystem activity, one must consider the situational variables that are present when a behavioral pattern is elicited, in order to examine the adaptiveness of the behavior to the environment. The analysis should include consideration of the following points.

situational: What stimulus conditions are present in the environment when the behavior is observed? What conditions does the individual describe as being important? What are the constraints of the situation? Who are the individuals in the setting, their roles and expectations? What factors serve as positive reinforcements of the behavior? What factors serve as negative reinforcements of the behavior?

goal: What apparent purpose does the behavior serve for the individual? What is the desired outcome? Which subsystem does this behavior most closely approximate? What aspect of the overall goal of the subsystem does the behavior meet? Why is the goal of the behavior assigned to this subsystem and not to another? Does the behavior of the individual apply to a long-term goal? Or to a short-term goal? If both goals are present, are they related to the same or different subsystems?

choice: What other behaviors could the individual use to meet the same goal? Are the behaviors adaptive to the situaation? If maladaptive, what behaviors would be more adaptive to the situation? What are the characteristics of the individual that limit the available choices?

set: What is the characteristic mood of the individual? What causes him to become aware of the stimulus conditions? How does the individual show that he is aware of these contingencies? Does the individual habitually respond to similar situations with the same behavior? If the same pattern of behavior is consistently utilized, why? Are there similarities in the various situations in which the same behavior is elicted?

interaction of components: Is there a clear relationship between the stated goal of the individual, the situation, the set, and the chosen behavior? Is the chosen behavior appropriate to the goal? Does the set of the individual result in misperception of the environmental contingencies?

Preliminary Analysis of the Interaction between Behavioral Subsystems

This phase of the analysis examines the nature of the interaction among the eight subsystems and determines the variation over 24 hours that is characteristic of an individual's behavior. Are certain behaviors dominant at various points of the day? Are each of the subsystems equally developed in terms of complexity and variety of choices? Does the activity in one subsystem complement, or compensate for, the activity of another? And will activity in one subsystem be likely to initiate activity in a second subsystem?

Certain behaviors can be predicted to occur with a relatively high probability on the basis of time of day. For example, restorative behaviors are often dominant in the evening. But in those cultures that prescribe a siesta (a period of relaxation following the noontime meal), the sociocultural custom facilitates activity in the restorative subsystem at a specified time each day—in the afternoon. This also illustrates the relationship between ingestive behaviors—eating the noontime meal—and subsequent restorative behaviors—napping. Eating also occurs at fairly regular times of day and enables predictions regarding the occurrence of this behavior. The more explicit the sociocultural rules and customs regarding distribution of activity throughout the day, the easier it is to examine the pattern of behavior of an individual in relationship to the general pattern for that society. Knowledge of an individual's occupation will also help in analysis of subsystem activity. A nurse who is employed in a hospital setting that requires periodic shift rotation will have a different pattern of 24-hour distribution of behavioral activity than will a person who works from eight to five every day, five days a week.

In order to examine the usual pattern of activity for an individual it is helpful to have him describe, in detail, a typical day. The task of the interviewer is to elicit as much specific detail as possible regarding the behaviors that may be present during this typical day. The ideal, of course, would be to follow that individual throughout the day and observe the presence or absence of reported behaviors, but this is relatively impractical. Within a hospital setting, however, the collaborative observations of the nursing staff on all three shifts could achieve a picture of the 24-hour distribution of behavior during hospitalization. This then could be contrasted to the reported distribution under nonhospital conditions for any areas of similarity or dissimilarity.

It is assumed that each of the eight subsystems is of equal importance to the ongoing function of the individual and as such should demonstrate equal development in terms of complexity and differentiation. Examination of the goals that affect an individual's behavior should reveal a fairly even distribution among the eight subsystems. However, one or more subsystems may be structurally insufficient—i.e., have a low level of activity, few and highly restricted choices, or a fixed set—and this insufficiency may be compensated for by increased levels of activity in another subsystem. For example, a person may report a total absence of behavioral actions related to the dependency subsystem, whereas there is a dominance of achievement behaviors present in his repertoire. It is possible that the achievement subsystem is compensating for the insufficiency of the dependency subsystem and that assistance in the development of help-seeking behaviors would restore balance between the two subsystems.

The boundaries between each of the eight subsystems of behavior are open to one another, and activity in one can initiate activity in another. The extent of this relationship varies, and some of the subsystems are more closely linked than others. For example, the process of ingesting food has a high probability of initiating activity in the eliminative subsystem through the reflex pathways mediated by the gastrointestinal system. Or, as a less obvious example, attending a social function with one's mate has a higher probability of eliciting subsequent sexual subsystem activity than of eliciting achievement behaviors. It is important to examine the nature of behavioral system activity for these subtle patternings of subsystem interaction because it represents the dynamic patterning of behavior that is characteristic of the individual and serves to differentiate him from all others.

Nursing Diagnosis and Implications for Nursing Intervention

The nursing diagnosis is a summary of the results of the foregoing analysis and describes the current level of behavioral system function. It shares with the medical diagnosis the function of serving as a prescription for subsequent intervention by the nursing team. The basic diagnosis should include statements on:

I. **Adaptive and functional behavioral system activity**

II. **Disturbance of ongoing behavioral system activity**
   1. Maladaptive behaviors as a consequence of situational constraints.
   2. Maladaptive behaviors as a consequence of alterations in biological, psychological, social, cultural, and environmental regulators.
   3. Disturbance in function of behavorial subsystem_____.

III. **Disturbance in behavioral subsystem interaction**
   1. Problem in the distribution of behavorial activity.
   2. Dominance of activity in subsystem_____.
   3. Insufficent activity in subsystem_____.

There are three basic forms of nursing intervention that can be used for one or all of these diagnostic statements. First, the nurse can *support* and *maintain* those behaviors that are adaptive to the constraints provided by the illness and the hospital setting, and are functional for the individual. This intervention assumes that the role of the nurse is to facilitate the use of personal modes of activity as much as possible while the patient is in the hospital. The second class of intervention relates to the role of the nurse as teacher. Maladaptive or malfunctional behavior may be the result of insufficient development of alternative choices, lack of exposure to different stimulus conditions, and a lack of personal committment to short-term and long-term goals. *Teaching* and *counseling* of the individual may provide him with additional ways of behaving that are more adaptive to his life style and preferences. A third approach to nursing intervention would be the active *modification* of behavioral patterns through the use of behavioral modification techniques. Such techniques can be helpful in assisting the individual to adapt to environmental contingencies that influence behavior. The interested reader should refer to Michael LeBow's text, *Behavior Modification: A Significant Method in Nursing Practice.*[4] This third approach would not be used if the disturbance is related to a temporary disruption of normal behavioral activity such as during an acute, short-term illness, or during a brief hospitalization. In such cases a combination of teaching an individual the necessary changes in behavior that facilitate adaptation to the environment and of support of those behavioral components that are adaptive would be the most appropriate approach.

SUMMARY.   The importance of behavioral assessment as a means of establishing a data base of information concerning the patient and his behavior prior to the onset of the illness cannot be overemphasized. An assessment protocol, organized on the basis of the behavioral systems analysis approach, has been presented for the reader. It is not expected that such an extensive assessment would be made for all patients. Rather, those portions of the tool that are related to a specific concern of the nurse, and that appear to be of greatest relevance for the care of the patient, should be used. A major advantage of organizing an assessment protocol in terms of a mode of nursing is that large amounts of data can be evaluated within the structure of an already existing framework.

Completion of the assessment is only the beginning of the process. The health team goals and patient goals can be identified and plans for attaining them can be developed and implemented. The behavioral systems approach, rather than depersonalizing the patient, facilitates the development of individualized plans of care. It enables a nurse to identify an individual's areas of strength and weakness and to predict possible effects of the experience of hospitalization and illness based upon a knowledge of the individual before his development of disease. As the one consistent, ever-present member of the health team, the nurse is in the unique position of being able to help the patient cope with his illness and develop adaptive patterns of behavior for the future. This can only be accomplished if the nurse is willing to spend the time and energy required to identify, through a comprehensive assessment, the factors that regulate and influence the individual's pattern of behavior. The potential result of such a process is that the nurse and patient will not pass one another as strangers in the collective mass, but rather that each will gain a unique experience as a result of having known the other. Perhaps the event of illness can then become a meaningful and positive occurrence in the lifetime of the individual.

STUDY QUESTIONS

1. Select a representative sample of items from the assessment protocol and administer the tool to a variety of individuals, both patients and nonpatients.

2. Based on the information obtained through the interview and through observations of current patterns of behavior, identify the adaptive and maladaptive behaviors for each person; make a nursing diagnosis.

3. Repeat the interview process using a nonstructured, open-ended approach. What are the differences in information obtained from a structured versus nonstructured interview?

This chapter presents two case studies of patients, based upon the behavioral system analysis approach. The analysis is in greater detail than would be necessary in a clinical setting to illustrate the process by which the adaptiveness and the functional status of behavior are identified. The first case study is the behavioral assessment of a woman admitted to a neurological medical unit for treatment of an acute myasthenic crisis; the second is a behavioral assessment of a male patient in a psychiatric outpatient setting.

Case One

Mrs. A. is a 33-year-old, married female, mother of three children, who was admitted on April 2, 1975 in an acute myasthenic crisis following a viral infection. This is her third admission to the hospital in the past four years. Mrs. A was in acute respiratory distress at the time of admission. She was unable to communicate verbally, but was in an apparent state of acute fear that she might die due to suffocation. She was cyanotic, diaphoretic, tachycardic, hypertensive, and dysphagic. Immediately following admission a tracheotomy was performed, which relieved her immediate respiratory distress. Her physical condition at the time of admission precluded an assessment interview until her condition had stabilized. The initial nursing plan of care was based on assessments from her two previous admissions, as well as on additional information provided by her husband.

GENERAL REGULATORS

Mrs. A. was born on July 4, 1942. She was an only child, was raised in a small town in the Midwest, and lived at home until she left for college. Her parents were active people whom she described as being leaders of the community. Her father was the local justice of the peace and a businessman; her mother was a housewife and periodically helped in their store. She was reared in the Methodist Church, but has not attended church since she left her parents' home in 1959. She describes herself as a "Christian who does not ascribe to one faith."

Mrs. A. has been married for seven years, a marriage that she describes as a very meaningful part of her life. She met her husband in college, and they married following her graduation with a degree in primary education. She taught in a local grade school until four years ago when the symptom of increasing fatigue, later diagnosed as myasthenia gravis, necessitated her resignation. She has three children, ages five, three, and two. The oldest is a boy and the other two are girls. She and her husband live in an upper middle-class neighborhood, in a single-dwelling residence that has five bedrooms and three baths. They have a live-in maid who was hired two years ago because of the increasing inability of Mrs. A. to maintain the household without assistance. The two older children attend school and nursery school, respectively, while the younger daughter is home in the care of the maid. Mr. A. is a successful attorney who was raised in a wealthy community, attended private schools, and has not expressed any financial restrictions that might prohibit the management of Mrs. A.'s illness.

Prior to the onset of the illness, Mrs. A. described herself as a very active, involved member of her family and community. She and her husband were involved in the Young Republicans, attended local concerts and plays on a frequent basis, at least three times a month, and entertained socially two to three times a month. During the summer, when she was not involved at her school, Mrs. A. volunteered as a counselor to underprivileged children two days a week, and went to the beach and to community recreational activities with her own children the remaining time. She enjoys reading, both fiction and nonfiction books, painting, needlework, and cooking. Before the onset of her illness she was a sport's enthusiast, playing golf and tennis on the weekends as often as possible. With the onset of the myasthenia gravis she has gradually had to reduce her participation in all of the foregoing activities, with the exception of reading, painting, needlework, and cooking. She is able to attend a party for a few hours, if she rests the entire day and does not involve herself in any other activities.

Mrs. A. described herself as a "well-put-together person" prior to her illness. She rarely encountered situations in which she felt uncomfortable. When she was required to do something that she did not particularly enjoy she found that it was most helpful to get it over with as soon as possible rather than to contemplate it. She would experience mild apprehension regarding new or unknown situations, such as when she would attend business functions of her husband's firm, but once she was at a function she was able to relax and enjoy herself. She rarely thought about illness or death, but since the onset of her illness she has had periods of becoming obsessed with the feeling that she may not live to see her children grow to adulthood.

Mrs. A. had been in remarkable health as a child, experiencing only the usual childhood diseases, ear infections, and skinned knees. Her first pregnancy was uncomplicated and her stay in the hospital for three days, following delivery of a healthy 8-pound male infant, was described as a "pleasant experience." She had enjoyed the feeling of being taken care of by the nursing staff, but was also glad to return home to her usual pattern of life. Her second hospitalization was for diagnostic purposes, at which time the diagnosis of myasthenia gravis was confirmed and she had a surgical removal of the thymus gland. This hospitalization was in 1971, and was the first at this institution. During the stay in the hospital she became quite depressed and was emotionally labile and tearful for extended periods of time. She was unable to express the content of her feelings. The depression eased following her surgery, although her mood was not elevated, nor was she optimistic regarding the future. Her second hospitalization at this hospital was her first myasthenic crisis, which occurred during her last pregnancy. She was in an extreme state of panic at the time of admission, unable to breathe. The crisis was resolved in two weeks and after stabilization on the medication she was discharged. During the hospitalization she frequently requested assistance from the nursing staff, was angry when she was requested to wait for awhile, and refused to remain in bed once her physical condition had stabilized. She refused to discuss the situation preceding the onset of the crisis and any possible changes in her life style that might assist her in adjusting to her limitations of energy. Her husband made the decision to hire a live-in maid during this hospitalization and implemented it without her approval.

BEHAVIORAL SUBSYSTEM ACTIVITY

Ingestive Subsystem

Prehospitalization patterns. Mrs. A. weighs 125 pounds, normal for her age, sex, and body build. She eats three meals a day in the following pattern: breakfast, upon rising at 9:00 A.M.; lunch at 1:00 P.M. and dinner at 8:30 P.M., upon her husband's arrival home from the office. She reported that the illness has required her to decrease the amount of food she normally ingests, because of the necessity for extended periods of rest and the consequent decrease in caloric requirements. She prefers all foods, with the exception of such organ meats as heart, brain, and liver. During acute exacerbations of the myasthenia gravis she experiences some difficulty in chewing and has had to modify her diet to soft and/or liquid foods.

She expressed a preference for coffee and milk with meals, except in the evening when she has a glass of wine with dinner. The illness has not had any direct effect on her pattern of ingesting fluids, to her knowledge. When teaching, she restricted drinking coffee to the morning hours, but since her life style has become more sedentary she drinks more coffee because of boredom.

Mrs. A. reported no restrictions related to the sensory modalities. She used to enjoy taking arts and crafts classes in order to learn new hobbies but has been unable to for the past two years. She reported that she reads the newspaper in depth each day and reads current news magazines, literary reviews, and women's periodicals in order to remain informed about the world and community.

Impact of hospitalization. Because of the respiratory distress it was necessary to perform a tracheotomy at the time of admission. Mrs. A. has a disturbance of the ingestive subsystem in the areas of food, fluid, and oxygen. She will require frequent tube feedings of a balanced, liquid diet. In order to maintain fluid balance, intravenous fluids will be administered throughout the time that she is unable to drink fluids by mouth. Mechanical assistance is also required to maintain an adequate oxygen supply, and careful monitoring and frequent suctioning of the airway is necessary to maintain patency.

Predicted patterns at time of discharge. Mrs. A. will probably be discharged with a permanent tracheotomy. During the hospitalization she will need to learn self-suctioning procedures and the technique for changing inner cannulas, so that she can be independent in her care. Her diet will be composed of soft, pureed foods which can be ingested with a minimum of chewing. It will also be necessary for her to learn to insert a nasogastric tube should tube feedings be required. Mrs. A. must also be counseled regarding the importance of eliminating alcohol from her diet; she can no longer ingest wine with dinner, because of the effects of alcohol on the respiratory system. It is expected that there will be no change in her preferred sources of reading.

Eliminative Subsystem

Prehospitalization patterns. Mrs. A. has had no difficulties with bowel and bladder control. She normally has one bowel movement per day; at infrequent intervals it is necessary for her to use some form of laxative in order to stimulate a bowel movement but this has not been a problem recently. She menstruates approximately every 28 days, and experiences minimal discomfort at the onset of menses. She reports that she felt relaxed and comfortable during her most recent pregnancy.

Mrs. A. has experienced periods of depression since confirmation of the diagnosis of myasthenia gravis, during which she cries easily and sits for hours doing nothing. During such times she does not feel like talking to anyone. These episodes last approximately four to five days and then, for no apparent reason, disappear. She begins to feel better, and is then able to talk about these feelings with her husband. She said that she prefers to be left alone by everyone when she feels sad. She experiences difficulty in the expression of anger, particularly toward her husband because she feels totally dependent upon their relationship. Her feelings of anger are expressed indirectly, through sarcastic comments, or "unreasonable" demands. Mrs. A. stated that she has little difficulty expressing feelings of happiness, "although I rarely feel that way in recent months."

Impact of hospitalization. At the time of admission, Mrs. A.

was experiencing a state of fear regarding whether or not she would survive the respiratory distress. Once the tracheotomy was performed and her breathing was supported by mechanical processes she visibly relaxed and was less agitated in her movements. The primary disturbance present at this time is an interference with her ability to communicate with others in her environment due to the tracheotomy; extreme diaphoresis due to the pathological process; and incontinence of feces and urine. Nursing measures should be immediately instituted to maintain her communication with others through writing. It is expected that she may experience a period of depression during the hospitalization which may interfere with her communication patterns with the staff.

Predicted patterns at time of discharge. By the time of discharge there should be no permanent changes required in eliminative behaviors with the exception of having learned to communicate with the tracheotomy.

Sexual Subsystem

Prehospitalization patterns. Mrs. A. felt that it was very important to her to be a "feminine" person, as expressed in her relationship to her husband and children. Femininity meant hidden strength and purpose, pliability, softness, and nurturing of the family. She described her husband as being very masculine as expressed in his logic, ability to control his emotions, and physical prowess. Her role as mother and wife of the family had been modified somewhat with the onset of her illness in that she was less active in the direct caring of the family members. Instead she fulfilled the role by directing the activities of others, such as the live-in maid.

Their frequency of intercourse was reported to have decreased since the onset of the illness, from an average of twice a week, to twice a month. The primary difficulty is that she has to anticipate when she and her husband would have intercourse, in order to rest throughout the day so that her energy reserves would be sufficient. This need to anticipate, i.e., adjustment of her preparatory set, interferes with her feelings of spontaneity and she has not enjoyed intercourse as a result.

Impact of hospitalization. The hospitalization restricts her activity in the sexual subsystem of behavior by virtue of the illness, and the rules of the system. She is not expected to experience any disturbance in her perception of herself as a woman, but may have difficulties adjusting to required changes in terms of her roles as wife and mother. Frequent visits by her husband and communications from her children will assist her in adjusting to these restrictions.

Predicted patterns at time of discharge. A return to prehospitalization patterns.

Affiliative Subsystem

Prehospitalization patterns. Prior to the onset of the illness Mrs. A. was an active woman, involved in community organizations and in her local political group. She had many friends in these groups with whom she and her husband had social evenings in addition to the meetings. They are members of the Methodist Church, but the primary reason for this membership is because of the social desirability for a bright attorney to be a member of a church. Mrs. A.'s family is still located in the Midwest. Mr. A's family is in the local community and they see each other on a regular basis, approximately once a week for dinner. Mrs. A. has a close, intimate relationship with her mother-in-law and she feels very supported by her husband's family.

Since the onset of her illness Mrs. A. has experienced a lack of interest in the organizations of which she had been a member, and has gradually cut off any contact with these groups. She stated that she would prefer not to have any contacts with these groups because she feels so different, and no longer a part of their activities. She has maintained personal contact with two women, who visit her frequently at home. She and her husband have not entertained friends for social evenings during the past year because of the extreme fatigue Mrs. A. experiences during such evenings. She denied feeling any sense of isolation from others, and is content to relate to her children, her husband, and his family.

Impact of hospitalization. During the hospitalization Mrs. A. will be restricted in the amount of time she can be in contact with the individuals who are most meaningful in her life at this time. She may need to be encouraged to relate to other patients and/or staff at those times when she would like to interact with others. Mrs. A. enjoys reading and watching television, which are substitute forms of affiliation that will conserve her energies.

Predicted patterns at time of discharge. A return to prehospitalization pattern of behavior.

Dependency Subsystem

Prehospitalization patterns. Mrs. A. expressed a great deal of resentment regarding the necessity to be dependent upon others, particularly for physical care. She realized that there were areas in which she required assistance at home, such as cleaning and cooking, but it was difficult for her to request the help of the maid. She had always prided herself as being an independent person, and equated this with maturity. Consequently she felt that she was being placed in the position of a child by having to depend upon others.

Her initial reaction to the increasing periods of fatigue was to examine what activities she could easily forego in favor of other, more highly preferred activities. She finally went to the doctor only when the fatigue was such that she was unable to be active beyond the first three hours of the day. During the first hospitalization, for her pregnancy, she enjoyed the care of the nurses because she felt pampered, and knew that it was her choice. Since she has required hospitalizations because of illness, she has resented the necessity to be dependent upon others and has responded with a great deal of anger. She has never considered suicide, because of the sustenance provided by her family. Only if she were deserted by her husband would she contemplate killing herself.

Impact of hospitalization. The severe respiratory distress and subsequent tracheotomy place Mrs. A. in a position of dependence on the respirator for maintenance of oxygen. Her initial response to the respirator was relief and gratitude for the presence of the assistance. As her hospitalization progresses, however, it is important to assist her to not become overly dependent upon the respirator but rather to utilize it only when required. Mrs. A. has expressed difficulties in the area of dependency on others for personal hygiene and care. During her previous hospitalization on this unit she insisted upon maintaining her self-care, even on those days when to do so depleted her energies. It would be helpful to design a program which maximizes her abilities to participate in care and which would help her learn a more adaptive mode for requesting and accepting assistance of others.

Predicted patterns at time of discharge. Mrs. A. will have learned new alternative behaviors for eliciting the assistance of her family and maid, when they are needed. She will also be independent, as much as possible, in her self-care of her permanent tracheotomy, and in insertion of the nasogastric tube, when these are required. It is important to assist her in utilizing the services of others in such a way that she will be able to fulfill her goals for independent functioning while she simultaneously acknowledges the very real limitations of her physical condition in this area.

Aggressive-Protective Subsystem

Prehospitalization patterns. Mrs. A. has experienced great difficulty accepting the future implications of her diagnosis. Her reaction has been to deny the physical limitations as much as possible, and only extreme fatigue has been sufficient to require her to modify her life style. This process of denial has prevented her from making realistic changes within her home and family living patterns that would be adaptive to her physical illness. For example, she exposed herself to her children while they had influenza and consequently created the conditions for her own viral infection and subsequent pulmonary distress. Even though she was aware of the necessity for avoiding any exposure to infectious diseases, particularly those with respiratory involvement, she has consistently failed to follow through on this information. The denial has two aspects: (1) a fear of becoming physically dependent upon others, and (2) a fear of death. Behaviorally, her denial has been expressed through actions that are nonadaptive to her limitations, such as remaining active to the point of exhaustion at times, failing to take her medication as prescribed, and exposing herself to illnesses of the family.

Mrs. A. states that before the onset of the illness she rarely experienced situations in which she felt threatened or unable to respond in some way that permitted her to maintain some degree of control. She and her husband rarely had arguments, and on her part this was a matter of being happy in the relationship, rather than of avoiding any confrontation. Since the illness she has felt some difficulties in this area, such as when she becomes annoyed by the family and is afraid to express that annoyance for fear they might turn away from her. Her periods of depression have been characterized by withdrawal from her family and a minimum of interaction. She reflected that she had the feeling that if someone were to try to talk to her when she was feeling sad, she would be able to do nothing but scream at them to leave her alone.

She has avoided, as much as possible, learning anything about her illness or prognosis, and stated that she prefers it to be that way until she feels able to handle the information. Her doctor has attempted to provide her with specific information at times· but she has left his office, whenever possible. Consequently it has been necessary for him to work closely with Mr. A. in making arrangements, such as the live-in maid, that will reduce the pressures upon Mrs. A.

Impact of hospitalization. The circumstances surrounding Mrs. A.'s hospitalization were such that her fears surrounding the areas of physical dependence and potential death were heightened. Her level of anxiety appeared to be severe, based upon the behaviors of agitation, restlessness, undirected activity, pupil dilation, pulse rate, and blood pressure readings, after respiratory function had been stabilized. Once the anxiety has been decreased to a manageable level, it would be helpful to explore the effect of information related to myasthenia gravis upon her use of denial regarding the impact of the disease. Denial of the pathological process is analogous to the first stage of the grief and mourning process—denial of an anticipated or real loss of a relationship. In the normal process of grieving, following the stage of denial, there would be a period of expressed anger that would be directed toward "healthy" persons in her environment. Expressing these feelings of anger and hurt may be in conflict with her feelings of having to be dependent on others. Assisting Mrs. A. to accept her diagnosis would make it possible to provide her with more adaptive and functional modes of protecting herself from potential threats in the environment. The nurse can be helpful in this process by understanding and supporting the movement of the individual through the stages of acceptance.

Predicted patterns at time of discharge. Mrs. A. will either have increased her acceptance of her illness and have learned adaptive coping mechanisms that can sustain her level of health present at discharge; or, there will be a return to the prehospitalization pattern of denial and a high probability of exacerbations.

Achievement Subsystem

Prehospitalization patterns. Before the onset of her illness, Mrs. A. experienced a great deal of success as a primary school teacher. This was a highly pleasurable occupation for her which she was unhappy to relinquish, even though she knew that she could not physically maintain the level of energy required by the job. Since that time she has experienced little sense of mastery over her environment, and feels that she has no alteratives to select from that would provide her with a feeling of achievement. Instead, she expresses the belief that she is totally at the mercy of others in the environment and that this is the reason she finds it so difficult to be placed in a position of dependency.

Impact of hospitalization. The hospital environment and dependence upon staff and mechanical life supports can further under-score Mrs. A.'s feelings of having no mastery over her environment. It would be helpful to explore her interests; for example she indi-cated an interest in needlework and painting. Perhaps creating her own canvases or needlework patterns would provide a feeling of achievement that would not be incompatible with her varying levels of available energy. Any activity would have to be one that could be performed with a minimum of effort, while providing a feeling of achievement and success.

Predicted patterns at time of discharge. Mrs. A. will have identified areas of interest that would permit her to engage in behav-iors that would create a feeling of achievement and mastery.

Restorative Subsystem

Prehospitalization patterns. Mrs. A. recalled that before the onset of the disease she had boundless energy, was able to function with only six hours of sleep each night, and did not require rest pe-riods during the day. She was able to concentrate on a single project for extended periods of time, e.g., four to five hours, but enjoyed a change of interest and activity for relaxation.

The initial symptom of her illness was increasing periods of fatigue following minimal exertion. She gradually was unable to remain active for more than a few hours without a sustained period of rest. She reported that her strength was greatest at the time of arising, but that with repeated use of her muscles throughout the day the weakness ensued. There was a partial return of strength following rest. The disease process typically involves the striated musculature of the limbs and/or bulbar areas. The course is charac-terized by spontaneous remissions and exacerbations. Mrs. A. related that there were times during the past four years when she had experi-enced an almost total return of strength, but that these periods had become shorter and less frequent and had not occurred in recent memory. She has been increasingly confined to her bed and/or chair for most of the waking hours; she is able to be actively engaged in minimal physical exertion for approximately 45 minutes with little difficulty during the day.

Impact of hospitalization. The prehospitalization pattern of extended rest periods and minimal physical exertion will be main-tained in the hospital. This is an adaptive and functional pattern for Mrs. A. in view of the limitations imposed by the disease.

Predicted patterns at time of discharge. Extended rest periods will continue to be required following hospitalization. Activities that require little use of striated muscles should be encouraged, to provide a feeling of change to her daily routine. Reading, watching television, conversation with visitors or over the telephone, writing, and doing puzzles are possible alternative activities.

INTERACTION OF BEHAVIORAL SUBSYSTEMS

The primary area of disturbance in the interaction among the behavioral subsystems is in the relationship between the aggressive-protective subsystem, achievement subsystem, and dependency subsystem. Mrs. A.'s denial of her illness—as exhibited by exposing herself to potential infections, refusal to modify her patterns of activity, and difficulties in dealing with the potential of death—affects the activity of the achievement and dependency subsystems as well. The denial creates a deficiency in achievement behaviors by restricting the available choices to those that will avoid a confrontation with her physical status. Furthermore, it creates difficulty in the acceptance of the required dependency upon others. Mrs. A.'s set in dependency situations results in use of the aggressive-protective behavior of refusal to request help; simultaneously she perceives herself to have no mastery of her environment, and adapts to these feelings of helplessness by using aggressive-protective behaviors. Consequently these three subsystems are functioning in such a way that the aggressive-protective subsystem dominates the ongoing activity as a compensation for insufficient activity in the dependency and achievement subsystems.

NURSING DIAGNOSIS.   The following summary of Mrs. A.'s current pattern of behavior is based on the preceding analysis.

1. Behavioral disturbance in the function of the *ingestive* subsystem, specifically related to the ingestion of food, fluids, and oxygen, as a consequence of the change in the pathological process.
2. Behavioral disturbance in the function of the *eliminative* subsystem, specifically related to the ability to communicate with others and to express internal feeling states.

3. Behavioral disturbance in the level of activity of the *sexual* and *affiliative* subsystems, as related to the constraints of the hospital setting. This is a temporary disturbance.

4. Behavioral disturbance in the structure and function of the *dependency* subsystem, specifically related to the components of goal, choice, and set. This disturbance is compensated for by increased activity in the aggressive-protective subsystem.

5. Behavioral disturbance in the structure and function of the *aggressive-protective* subsystem, specifically related to the components of choice and action, in the use of denial, refusal to discuss the implications of her physical condition, and failure to avoid potential infections.

6. Behavioral disturbance in the structure and function of the *achievement* subsystem, specifically a deficit in available choices and level of activity.

7. Behavioral adaptation and function in the *restorative* subsystem, which needs to be maintained throughout the hospitalization.

8. The pattern of interaction among the eight subsystems of behavior indicates that the restorative and aggressive-protective behaviors are dominant; dependency and achievement behaviors are insufficient; and the remaining four subsystems of behavior demonstrate moderate levels of activity.

9. All subsystems of behavior have been modified in structure, function, and level of activity by the impact of the regulator of illness.

Case Two

The presentation of the second case differs in format from the preceding case because of the characteristics of the clinical setting. A psychiatric outpatient client seeks the assistance of a health professional because of the perception by himself, or by others in his environment, that there is a disturbance in his current pattern of behavior. The focus of the nurse practitioner, or counselor, is to help the client identify and modify these areas of maladaptive or malfunctional behavior, while he remains in the community setting. The client is assumed to bring to the therapy session his healthy and nonhealthy patterns of relating to the world and the environment, so that the behavior observed in the clinical interaction is representative of his overall behavior. Consequently, the behavioral analysis and assessment are concerned with a wide variety of situational and behavioral factors and the ways in which the individual responds to these situations in his life, and not with the impact of therapy per se.

Mr. T. is a 23-year-old, single, Caucasian male, who was seen as a walk-in, self-referral in the emergency room of a psychiatric hospital. He stated that the reason for his visit was his concern over his increasing inability to control his impulsive behavior. During a fight with K., his girlfriend, the evening before, he impulsively went to the bathroom and took a double-edged razor blade and with complete coolness and lack of passion slit his right wrist in her presence. Once he had penetrated the skin, he commented that he "suddenly woke up" and became aware of what he had done. He went to the emergency room of the local hospital, accompanied by his girlfriend, where they told the surgeon on duty that he had slipped and fallen through the window of her apartment. While Mr. T. was giving this information, he was holding his right wrist, which was bandaged. He had a nervous laugh, and frequently prefaced his comments by saying, "I know it's silly but . . . . " He was neatly groomed, dressed in neat, casual clothes; his posture was slightly strained as though he were prepared for flight at any moment. He never removed his hand from his right wrist throughout the entire interview.

GENERAL REGULATORS

Mr. T. is an only child of his natural parents. When he was twelve, his mother committed suicide during a quarrel with his father. His father later remarried, when Mr. T. was fourteen; his step-mother is "not at all like the ones in fairy tales." She has a daughter, two years younger than Mr. T., who moved in with him and his father. He described their relationship as one of mutual intolerance, punctuated by many conflicts.

Mr. T. was born and raised in a large metropolitan city. He has traveled throughout the southwest and expressed an interest in being free to roam the world; the only problem is that it requires money. He attended public schools and was a "B" student at the time of graduation from high school in 1970. He completed a semester at college but withdrew because his father failed to provide him the promised economic support. Since that time he has "bummed around" with friends and has worked, periodically, as a laborer on a construction gang. At the time of the interview he was employed as a parking lot attendant. His father is a professor of astronomy at the state university and his mother had graduated from college with honors in English. He felt that there was an extreme emphasis on going to college as being the "right thing to do," but found himself extremely resentful of his father's lack of support.

He is not married, and has no interest in developing a permanent relationship with any woman. He has been dating his current girlfriend for the past year and a half. They lived together for three months, but then decided to maintain separate residences because of their frequent quarrels. Since that time he has had dates "on the side" with other girls; he is positive that K. is unaware of this behavior. He shares an apartment with a male friend with whom there is a mutual understanding that if one wants to have exclusive use of the apartment for an evening, the other will gracefully exit. The apartment is located in a low-income area of the community. Mr. T. described it as just barely standing up amid all of the garbage and weeds that surround it. He does not like to have any of his family visit him there because it makes him uncomfortable.

Mr. T. was raised in the Jewish faith as a child, and was bar mitzvah at the age of thirteen. He has not attended temple since that time, and said rather cynically that the only reason he went through the ceremony was for the money and pens that all his relatives sent him. His father observes the holidays but attends temple only sporadically. Mrs. T., his stepmother, is a Roman Catholic who has left the church. Mr. T. said that his philosophy of life is that the only way to get ahead is to outfox all the other people. "You have to know how to manipulate the system; everyone does it. You have to be silly if you don't."

Mr. T. stated that he experienced difficulty in responding to situations that involved some degree of reluctance on his part. He would first attempt to handle it by force, or would avoid dealing with it if at all possible. If there was absolutely no way of avoiding the situation he then would try to make it as favorable to him as possible. If this required manipulation of other people in the setting that was unfortunate for those people. He denied feeling anxious in any specific situations; instead he stated that he feels very calm and controlled in all situations, with the exception of interactions with his girlfriend, K.

He has never been hospitalized for medical or psychiatric reasons. However, two years before this interview, Mr. T. was found to have a duodenal ulcer, believed by his physician to be related to a psychological stress condition. He was prescribed Donnatol and an antacid preparation for symptomatic relief. Mr. T. recalled that during this period in his life he was under pressure from his father to continue attending college. Mr. T. was unable to tell his father that he was angry with him for not having supported his education, but instead chose to remain silent and refused to speak to his father unless it was absolutely imperative. He said that in this respect he was similar to his father, who would have extended periods of controlled silence when he became angry or upset. Other than that shared characteristic, Mr. T. denied any physical or psychological characteristics in common with his father.

Mr. T. stated that for him the most important thing in life is to be on top of everything that happens to him. When asked whether he agreed with the statement that the important thing is not whether you win or lose, but how you play the game, he laughed, looked skeptical, and said "Come on."

BEHAVIORAL SUBSYSTEM ACTIVITY

Ingestive Subsystem

Mr. T. is a well-nourished young man of 23. He reported that he eats three meals each day, one of which is composed primarily of snack foods. His main meals each day are breakfast and dinner, at which time he said that he tries to eat a well-balanced, nutritious diet. He prefers vegetables, fruits, and milk and tries to eat as little meat as possible. One year ago he ate a vegetarian diet for six months to see whether it would help his ulcer symptoms and felt that there was some improvement. He denied the use of drugs other than antacids and Donnatol; he does not smoke and abstains from alcohol.

He stated that he enjoys reading for the information that he can derive from it and reads a wide variety of subjects, ranging from biographical studies of historical figures, to classical poetry, such as Dante's *Inferno*. He has no known sensory limitations, e.g., he does not wear glasses and appeared to comprehend questions readily. Mr. T. said that he prefers a relatively quiet environment, although he can concentrate under noisy conditions.

Eliminative Subsystem

Mr. T. reported normal eliminative function related to his bowels and bladder. He usually has one bowel movement a day, and does not remember ever having had to use a bowel stimulant. He admitted having difficulty expressing his feelings, particularly those of annoyance and anger. He stated that when he gets annoyed at a person he doesn't know, a salesperson or counter girl at a restaurant, he will leave the situation rather than express his annoyance to the individual. The usual stimulus that triggers this annoyance is his feeling that someone is either "dumb" or excessively slow. With friends he will try to laugh off his annoyance; in respect to K., his girlfriend, he finds himself beating her when he becomes annoyed with her. This happens approximately two times a month, and is usually related to his feeling that she deliberately misunderstands what he says to her.

Mr. T. stated that most of his friends view him as a happy-go-lucky person and he goes out of his way to confirm this image. He said that people really don't know him because he won't let them see who he really is. In respect to the expression of feelings other than anger, he related that he hides them all behind a clown-like facade. Throughout the entire interview, Mr. T. maintained a very controlled monotonous tone to his verbalizations, except when he expressed his concern regarding his loss of control during arguments with K.

Sexual Subsystem

Mr. T. reported that he views himself as a normal male, with the "usual" sexual desires of someone his age. When asked to explain what he meant by "usual," he became somewhat hesitant, blushed, and laughed, saying "Oh, you know what's normal." He stated that he engages in intercourse three to four times a week and has experienced no problems in this area. The concept of masculinity is very important to him. He was raised to believe that men are superior to women, and that they are the dominant sex. He described K. as a very "feminine" woman, i.e., "submissive, gentle, and compliant." He has no intentions of marrying now, although he would not be opposed to establishing a long-term relationship with K. Mr. T. definitely ruled out any interest in becoming a parent at this point in his life. He stated that he is not ready to be tied down by children and a domestic scene.

Affiliative Subsystem

Mr. T. currently has limited contact with his family. He stated that every time they were together, there would be a blowup over something that he had, or had not done. He feels that he has nothing in common with his father and does not care to see him unless it is absolutely unavoidable. His primary relationship at this time is with K. He described it as being primarily sexual on his part; when he wants to talk to someone he would prefer doing so with some of his buddies. The qualities of those persons he most admires relate to being in control of events and having the ability to manipulate situations to effect a desired outcome. He likes to spend at least three to four evenings a week out with the guys.

Mr. T. belongs to no organizations and expressed no interest in doing so. He felt that the effect of organizations is to force everyone into acting and believing in the same way. He stated that his parents had always been involved in a lot of "do-good" groups and that he had absolutely no use for them. He no longer is affiliated with the temple and expressed the view that religion is the "narcotic of the masses."

Dependency Subsystem

Mr. T. described himself as being totally independent from all others, both physically and psychologically. He said that there was actually no one person in his life who was important enough to him to make him change his values. When asked what he does when he needs the help of an expert in some area in which he is not knowledgeable, he initially denied that there were any such situations in his life. After further questions, he finally admitted that if he were unable to do something on his own, he would first seek help from a friend and then, if necessary, from an expert. He indicated that he tries to avoid new situations as much as possible.

Mr. T. felt that his suicide attempt of the evening before had been a bid for attention from K. He said that although it was uncalculated, that it must have been designed to get her attention, otherwise he would have waited until she left to slash his wrist. Philosophically he did not feel that suicide was a reasonable solution to the problems of his life, but he was unsure what the appropriate solution might be. When asked whether he felt that psychotherapy might be a possible solution he indicated that he was not sure.

Aggressive-Protective Subsystem

The characteristic way in which Mr. T. protects himself from potential threat is to avoid confrontation. He stated that he avoided direct confrontation with others because of his concern that he might lose his temper and end up in a physical battle. He described himself as a coward in most situations, although his friends would be surprised to hear his description. With the exception of his relationship with K., Mr. T. could not recall any situation that involved an element of threat with which he had dealt in any other way than by leaving the scene as quickly as possible.

Achievement Subsystem

Mr. T. expressed a great deal of dissatisfaction with his present level of achievement but has been unable to make a plan of action for meeting the necessary requirements for an occupation that would provide him with a feeling of success. He had no specific major in mind when he attended the state college, and even at the time of the interview had extremely vague plans for his future.

He stated that he does not enjoy competitive situations, and has observed that his performance deteriorates under such conditions. He is now employed as a parking lot attendant and stated that he enjoyed nothing about the job, except that he is outdoors. He is not interested in mechanics, nor in artistic endeavors. As a child he wanted to become a fireman, but his father strongly opposed that choice as being unacceptable for his son. Mr. T. is of above average intelligence, but his feelings of pressure in this area interfered with his performance on the Wechsler Adult Intelligence Scale. He explained the reasons for errors on the arithmetic subtests by stating that if he had a calculator he would be able to give the right answer.

Restorative Subsystem

Mr. T. stated that he has no difficulty in the areas of rest and sleep. He sleeps eight hours each night and awakens with a feeling of being refreshed in the morning. During the day he takes a brief nap following lunch, if it is possible. He enjoys physical activities, such as swimming, running, and hiking. During the winter he enjoys skiing, although his lack of funds has curtailed this activity. His last vacation was at the age of 15, although he indicated that whenever he feels like taking some time off for himself he just quits his job and bums around until it is necessary for him to return to work.

ANALYSIS OF SUBSYSTEM FUNCTION
AND ADAPTIVENESS

The preliminary analysis of the assessment information obtained from Mr. T. indicates that there is a disturbance in the function of the eliminative, affiliative, dependency, aggressive-protective, and achievement subsystems of behavior. The primary disturbance in each of these subsystems is a lack of a wide variety of alternative choices for selecting behaviors in the various situations which Mr. T. encounters each day. Instead his reported behaviors tend to be highly rigid and nonspecific. For example, all situations in which he perceives a potential or real threat are responded to by the behavior of avoidance; flight appears to be the only option available to him at this time. The choices available to him for meeting the goals of the achievement subsystem are particularly limited; he was unable to verbalize one specific area in which he felt that he could attain a feeling of success and mastery with the possible exception of manipulating people and "the system." The dependency subsystem of behavior is characterized by a rigid set to exclude perception of any situation as requiring the assistance of others. The affiliative subsystem demonstrates a limited availability of choices for Mr. T. to gain a feeling of relatedness. In general the basic area of interpersonal relationships is involved in each of these behavioral disturbances. The inability of Mr. T. to relate to others on any successful level appears to be a dominant feature of his behavior at this time. Finally, the eliminative subsystem disturbance, which involves insufficient action in the expression of feeling states, has resulted in the development of a physical disturbance, a duodenal ulcer, which may be viewed as a process of "being eaten up" by one's feelings.

The ingestive, sexual, and restorative subsystems appear to be adaptive and functional in terms of behavior at this time. The sexual subsystem is the only behavioral subsystem that involves an element of interpersonal relationships which demonstrates a relative lack of disturbance. It is possible that the activity in this subsystem has compensated for the disturbances in the others.

NURSING DIAGNOSIS. There is behavioral system disturbance, specifically expressed in the activity of the eliminative, affiliative, dependency, aggressive-protective, and achievement subsystems of behavior. The specific disturbance in the set of dependency subsystem may interfere with the establishment of an effective therapeutic relationship; Mr. T. must perceive himself as requiring assistance before he can become engaged in therapy.

Footnotes

preface

1. Dorothy E. Johnson, "One Conceptual Model of Nursing," paper presented at Vanderbilt University, Nashville, Tennessee, April 25, 1968.

one

1. World Health Organization, "Constitution of the World Health Organization," *Public Health Reports* 61 (1946): 1268–77.
2. Ruth Wu, *Behavior and Illness* (Englewood Cliffs, N.J.: Prentice-Hall, Inc., 1973), pp. 75–87.
3. Dorothy E. Johnson, "One Conceptual Model of Nursing," paper presented at Vanderbilt University, Nashville, Tennessee, April 25, 1968.
4. Charles Tart, "Altered State of Consciousness," paper presented at UCLA Neuropsychiatric Institute, Los Angeles, California, December 4, 1972.
5. Paul Ekman, Wallace V. Friesen, and Phoebe Ellsworth, *Emotion in the Human Face: Guidelines for Research and an Integration of Findings* (New York: Pergamon Press, Inc., 1972).
6. George Orwell, *1984* (New York: Harcourt Brace Jovanovich, 1949).
7. California Mental Health Act, enacted January, 1967. Welfare and Institutions Code: Section 5000–5699.
8. Michael D. LeBow, *Behavior Modification: A Significant Method in Nursing Practice* (Englewood Cliffs, N.J.: Prentice-Hall, Inc., 1973).
9. Johnson, "One Conceptual Model of Nursing."

two

1. Ludwig von Bertalanffy, "General System Theory and Psychiatry," in *American Handbook of Psychiatry*, III, ed. S. Arieti (New York: Basic Books Inc., 1966): 705–21.
2. James G. Miller, "Living Systems: Basic Concepts," in *General Systems Theory and Psychiatry*, ed. W. Gray, F. Duhl, and N. Rizzo (Boston: Little, Brown, Co., 1969).
3. Joseph Notterman and R. Trumbull, "Note on Self-regulating Systems and Stress," *Behavioral Science* 4 (1950): 324–27.

three

1. Daryl Bem and Andrea Allen, "On Predicting Some of the People Some of the Time," *Psychological Review* 81 (1974): 506–20.
2. Dorothy E. Johnson, "One Conceptual Model of Nursing," paper presented at Vanderbilt University, Nashville, Tennessee, April 25, 1968.
3. Ibid.
4. Rudolph Moos, "Situational Analysis of a Therapeutic Community Millieu," *Journal of Abnormal Psychology* 73 (1968): 49–61.
5. Walter Mischel, "Toward a Cognitive Social Learning Reconceptualization of Personality," *Psychological Review* 80 (1973): 252–83.

four

1. Dorothy E. Johnson, "One Conceptual Model of Nursing," paper presented at Vanderbilt University, Nashville, Tennessee, April, 25, 1968.
2. Irvin Child, "Socialization" in *Handbook of Social Psychology*, Volume II, ed. Gardner Lindzey (Reading, Mass.: Addison-Wesley Publishing Co., 1964) p. 661.
3. Sigmund Freud, *The Psychopathology of Everyday Life*, in *The Standard Edition of the Complete Psychological Works of Sigmund Freud,* trans. James Strachey (London: Hogarth Press, 1901), pp. 61–64.
4. Carl G. Jung, *Studies in Word Associations*, Vol. II of *Collected Works,* trans. Leopold Stein (Princeton, N.J.: Bollingen Series, Princeton University Press, 1973).
5. Franz Alexander, *Psychosomatic Medicine* (New York: Norton, Inc., 1950).
6. Paul MacLean, "Psychosomatic Disease and the 'Visceral Brain': Recent Developments Bearing on the Papez Theory of Emotion," *Psychosomatic Medicine* 11 (1949): 338.
7. B.G. Rosenberg and Brian Sutton-Smith, *Sex and Identity* (New York: Holt, Rinehart and Winston, Inc., 1972) p. 80.
8. Alex Comfort, *The Joy of Sex* (New York: Simon and Schuster, Co., 1972).
9. *"J" The Sensuous Woman* (New York: Dell Publishing Co., 1969).
10. David Reubin, *Everything You Always Wanted to Know about Sex . . . But Were Afraid to Ask* (New York: David McKay Co., 1969).
11. Zick Rubin, "Lovers and Other Strangers: The Development of Intimacy in Encounters and Relationships," *American Scientist* 62 (1974): 182–90.
12. Robert Alberti and Michael L. Emmons, *Your Perfect Right*, 2nd ed. San Luis Obispo, Calif.: Impact) 1974.
13. William Dement, "The Effect of Dream Deprivation," *Science* 131 (1960): 1705–07.

14. Laverne C. Johnson, "Are Stages of Sleep Related to Waking Behavior?" *American Scientist* 61 (1973): 326–38.
15. Jerome L. Singer, "Daydreaming and the Stream of Thought," *American Scientist* 62 (1974): 417–25.

five

1. S.W. Tromp, "Weather, Climate and Man," in *Handbook of Psysiology: Adaptation to the Environment* ed. D. Dill, E. Adolph, and C. Wilber (Washington D.C.: American Physiological Society, 1964) pp. 283–93.
2. Narda Trout, *Los Angeles Times*, August 12, 1971.
3. W. Griffitt, "Environmental Effects on Interpersonal Affective Behavior: Ambient Effective Temperature and Attraction," *Journal of Personality and Social Psychology* (July 1970): 240–44.
4. An international candle is the total luminous energy emitted in all directions by a standard candle with a flame'one inch high.
5. A millimicron (mμ) equals 1/1,000,000 millimeter.
6. A decibel is 1/10 of a bel which is a ratio scale established by reference to a tone of approximately 10^{-16} watts per cm², the absolute threshold for a 1,000 cps tone.
7. Edgar D. Adrian, "The Basis of Sensation: Some Recent Studies of Olfaction," *British Medical Journal* Part 1 (1954): 289.
8. David Krech and Richard S. Crutchfield, *Elements of Psychology* (New York: Alfred A. Knopf, Inc., 1962) pp. 68–69.
9. Alfred Baldwin and Clara Baldwin, "The Study of Mother-Child Interactions," *American Scientist* 61 (1973): 714–21.
10. Rudoph Moos, *Ward Atmosphere Scale Manual* (Palo Alto, Calif.: Consulting Psychologists Press, 1974).
11. John B. Calhoun, reported by Daniel Rice, *Health Services World* 8 (1973): 3.
12. Johannes Müller, *Handbuch der Physiologie des Menschen für Vorlesungen*, 2 Vols. (Coblenz: J. Holsher, 1834–40), translated selection in R. Herrnstein and E. Boring (Eds.) *A Source Book in the History of Psychology* (Cambridge, Mass.: Harvard University Press) 1966, p. 26–33.
13. E.N. Sokolov, "Neuronal Models and the Orienting Reflex," in *The Central Nervous System and Behavior*, ed. Mary Brazier (New York: Joseph Macy, Jr. Foundation, 1960) pp. 187–271.
14. John I. Lacey, "Psychophysiological Approaches to the Evaluation of Psychotherapeutic Process and Outcome," in *Research in Psychotherapy*, ed. E. A. Rubenstein and M.B. Parloff (Washington, D.C.: American Psychological Association, 1959) pp. 160–208.
15. William L. Libby, Jr., Beatrice C. Lacey, and John I. Lacey, "Pupillary and Cardiac Activity during Visual Attention," *Psychophysiology* 10 (1973): 270–94.
16. R. Hernandez-Peon, H. Scherrer, and N. Jouvet, "Modification of Electric Activity in Cochlear Nucleus during 'Attention' in Unanesthetized Cats," *Science* 123 (1956): 331.
17. Jean Piaget and Bärbel Inhelder, *The Psychology of the Child* (New York: Basic Books, Inc., 1969).
18. M. Levine, "Human Discrimination Learning: the Subset-Sampling Assumption," *Psychological Bulletin* 74 (1970): 397–404.
19. Michael S. Gazzinaga, "One Brain—Two Minds," *American Scientist* 60 (1972): 311–17.
20. Ibid., p. 315.

21. R.A. Filbey and Michael Gazzinaga, "Splitting the Normal Brain with Reaction Time," *Psychonomic Science* 17 (1969): 335–36.

22. William James, "What is Emotion?," *Mind* 19 (1884): 188.

23. Carl Lange, *The Emotions,* trans. by I. Haupt in *The Emotions,* ed. K. Dunlap (Baltimore: Williams and Wilkins, Co. 1922).

24. Walter Cannon, "The James-Lange Theory of Emotions. A Critical Examination and an Alternative Theory," *American Journal of Psychology* 39 (1929): 106.

25. Phillip Bard, "A Diencephalic Mechanism for the Expression of Rage with Special Reference to the Sympathetic Nervous System," *American Journal of Physiology,* 84 (1928): 490.

26. S. Schacter and Jerome E. Singer, "Cognitive, Social, and Physiological Determinants of Emotional States," *Psychological Review* 69 (1962): 379.

27. A. Petrie, *Individuality in Pain and Suffering* (Chicago: University of Chicago Press, 1967).

28. Robert Roessler, "Personality, Physiology and Performance," *Psychophysiology* 10 (1973): 315.

29. Michael Goldstein et al., "Coping Style as a Factor in Psychophysiological Response to a Tension Arousing Film," *Journal of Personality and Social Psychology* 1 (1965): 290.

30. Michael Goldstein and J. N. Adams, "Coping Style and Behavioral Response to Stress," *Journal of Experimental Research and Personality* 2 (1967): 239.

31. Rieva D. DeLong, "Individual Differences in Patterns of Anxiety Arousal, Stress Relevant Information and Recovery from Surgery," (Doctoral Dissertation, University of California, Los Angeles, 1970).

32. Walter Cannon, *The Wisdom of the Body* (New York: Norton Co., 1932).

33. Marion A. Wenger, F. Nowell Jones, and Margaret Jones, *Physiological Psychology* (New York: Holt, Rinehart and Winston Co., 1956).

34. John I. Lacey, "Principles of Autonomic-Response Stereotypy," *American Journal of Psychiatry* 71 (1958): 50.

35. Bernard Engel and A. Bickford, "Response Specificity, Stimulus Response and Individual Response Specificity in Essential Hypertensives," *Archives of General Psychiatry* 5 (1961): 478.

six

1. David Krech and Richard S. Crutchfield, *Elements of Psychology* (New York: Alfred A. Knopf, Inc., 1962) p. 564.

2. Halbert B. Robinson and Nancy M. Robinson, *The Mentally-Retarded Child: A Psychological Approach* (New York: McGraw-Hill Book Co., 1965) pp. 76–77.

3. N. Juel-Nielsen, *Individual and Environment* (Copenhagen: Munksgaard, 1969) p. 1 ff.

4. Gardner Lindzey, "Behavior Genetics" in *Annual Review of Psychology,* vol. 22 Palo Alto, Calif.: Annual Reviews, Inc., 1971).

5. C. Burt and M. Howard, "The Multifactorial Theory of Inheritance and Its Application to Intelligence," *British Journal of Statistical Psychology* 9 (1956): 95.

6. Herbert A. Sprigle, "Who Wants to Live on Sesame Steet?" *Young Children* 28 (1972): 91–109.

7. Louis L. Thurstone, *Primary Mental Abilities* (Chicago: University of Chicago Press, 1938).

8. Nathaniel Kleitman, *Sleep and Wakefulness* (Chicago: University of Chicago Press, 1963).

9. T. Hellbrügge, "Development of Circadian Rhythms in Infants," *Cold Spring Harbor Symposium on Quantitative Biology* (Cold Spring Harbor, L.I., N.Y.: Biological Laboratory) 25 (1960): pp. 311–23.

10. J. Aschoff, M. Fatranská, and H. Giedke, "Human Circadian Rhythms in Continuous Darkness: Entrainment by Social Cues," *Science* 171 (1971): 213–15.

11. Mary Lobban, "The Entrainment of Circadian Rhythms in Man" *Cold Spring Harbor Symposium on Quantitative Biology* (Cold Spring Harbor, L.I., N.Y.: The Biological Laboratory) 25 (1960): 325–32.

12. Ross A. McFarland, "Air Travel Across Time Zones," *American Scientist* 63 (1975): 23–30.

13. Charles P. Richter, *Biological Clocks in Medicine and Psychiatry* (Springfield, Ill.: Charles C. Thomas, 1965).

14. S.M. Dornbusch, "*Afterword*," in *The Development of Sex Differences*, ed. Eleanor E. Maccoby (Stanford: Stanford University Press, 1966), pp. 204–22.

15. Eleanor E. Maccoby and Carol N. Jacklin, *The Psychology of Sex Differences* (Stanford: Stanford University Press, 1974).

16. Matina S. Horner, "Fail: Bright Women," *Psychology Today* 3 (1969): 36.

17. Katherina Dalton, *The Premenstrual Syndrome* (Springfield, Ill.: Charles C. Thomas, 1964).

18. Arnold Mandell and Mary Mandell, "Suicide and the Menstrual Cycle," *Journal of the American Medical Association* 200 (1967): 792.

19. H. Oleck, "Legal Aspects of Premenstrual Tension," *International Record of Medicine* 166 (1953): 492.

20. Therese Benedek and B. Rubenstein, "The Correlations between Ovarian Activity and Psychodynamic Process: I. The Ovulative Phase," *Psychosomatic Medicine* 1 (1939): 245.

21. Jeanine R. Auger, "A Psychophysiological Study of the Normal Menstrual Cycle and of Some Possible Effects of Oral Contraceptives" (Doctoral Dissertation, University of California, Los Angeles, 1967).

22. Rudolph H. Moos et al., "Fluctuations in Symptoms and Moods During the Menstrual Cycle," *Journal of Psychosomatic Research* 13 (1969): 37.

23. Peter Suedfeld, *Attitude Change: The Competing Views* (Chicago: Aldine-Atherton, Inc., 1971).

24. Leon Festinger, *A Theory of Cognitive Dissonance* (Evanston, Ill.: Row, Peterson, Inc., 1957).

25. Julian Rotter, "The Role of the Psychological Situation in Determining the Direction of Human Behavior," *Nebraska Symposium on Motivation* (Lincoln, Nebraska: University of Nebraska Press, 1955) p. 245.

26. Marvin Seeman, "On the Meaning of Alienation," *American Sociological Review* 24 (1959): 783.

seven

1. J. Marshall and S. Feeney, "Structured versus Intuitive Intake Interview," *Nursing Research* 21 (1972): 769.

2. Walter Mischel, *Personality and Assessment* (New York: John Wiley and Sons, Inc, 1968) pp. 235–80.

3. Agnes Wilkinson, "Assessment of Symptoms," *British Journal of Medical Psychology* 45 (1972): 3.

4. Michael LeBow, *Behavior Modification: A Significant Method in Nursing Practice* (Englewood Cliffs, N.J.: Prentice-Hall, 1973).

References

one

AL-ISSA, IHSAN, AND WAYNE DENNIS. *Cross-cultural Studies of Behavior.* New York: Holt, Rinehart and Winston, Inc., 1970.

BLOOM, BENJAMIN. *Stability and Change in Human Characteristics.* New York: John Wiley and Sons, Inc., 1964.

CHEIN, ISIDOR. *The Science of Behavior and the Image of Man.* New York: Basic Books, Inc., 1972.

DUFF, RAYMOND, AND AUGUST B. HOLLINGSHEAD. *Sickness and Society.* New York: Harper & Row, 1968.

KLUCKHOHN, CLYDE, HENRY A. MURRAY, AND DAVID M. SCHNEIDER. *Personality in Nature, Society and Culture.* New York: Alfred A. Knopf, 1969.

KRECH, DAVID, RICHARD CRUTCHFIELD, AND EGERTON BALLACHEY. *Individual in Society.* New York: McGraw-Hill Inc., 1962.

LINTON, RALPH. *The Cultural Background of Personality.* New York: Appleton-Century-Crofts, 1945.

MEAD, MARGARET. *Culture, Health and Disease.* London: Tavistock Publishers, 1966.

WU, RUTH. *Behavior and Illness.* Englewood Cliffs, N.J.: Prentice-Hall, Inc., 1973.

two

ACKOFF, RUSSELL. "Games, Decisions, and Organizations." *General Systems* 4 (1959): 145.

ALLPORT, GORDON. "The Open System in Personality Theory." *Journal of Abnormal and Social Psychology* 61 (1960): 301–11.

CHIN, ROBERT. "The Utility of System Models and Developmental Models for Practitioners." In *The Planning of Change,* 2nd ed., eds. Warren Bennis, Kenneth Benne, and Robert Chin, pp. 297–312. New York: Holt, Rinehart and Winston, Inc., 1969.

GRAY, WILLIAM, FREDERICK J. DUHL, AND NICHOLAS D. RIZZO, eds. *General Systems Theory and Psychiatry.* Boston: Little, Brown and Co., 1969.

HALL, A.D., AND R.E. FAGEN. "Definition of System." In *Modern Systems Research for the Behavioral Scientist,* ed. W. Buckley, pp. 81–92. Chicago: Aldine Publishing Co., 1968.

MILLER, JAMES G. "Living Systems: Basic Concepts." In *General Systems Theory and Psychiatry.* ed. W. Gray, F.J. Duhl, and N.D. Rizzo, pp. 51–133. Boston: Little, Brown and Co., 1969.

MINUCHIN, SALVADOR. *Families and Family Therapy.* Cambridge, Mass.: Harvard University Press, 1974.

NOTTERMAN, JOSEPH, AND R. TRUMBULL. "Note on Self-Regulating Systems and Stress." *Behavioral Science* 4 (1950): 324–27.

REINER, JOHN M. *The Organism as an Adaptive Control System.* Englewood Cliffs, N.J.: Prentice-Hall, Inc., 1968.

RUESCH, JURGEN. "A General Systems Theory Based on Human Communication." *General Systems Theory and Psychiatry.* ed. W. Gray, F.J. Duhl, and N.D. Rizzo. Boston: Little, Brown and Co., 1969.

SCHRODINGER, ERWIN. "Order, Disorder, Entropy." In *Modern Systems Research for the Behavioral Scientist.* ed. W. Buckley, pp. 143–46. Chicago: Aldine Publishing Co., 1968.

SLACK, CHARLES. "Feedback Theory and the Reflex Arc Concept." *Psychological Review* 62 (1955): 263–67.

VON BERTALANFFY, LUDWIG. "General System Theory and Psychiatry." In *American Handbook of Psychiatry,* Volume III, ed. S. Arieti, pp. 705–21. New York: Basic Books, Inc., 1966.

WEINER, NORMAN. *Cybernetics.* New York: John Wiley and Sons, Inc., 1948.

three

ALLPORT, GORDON W. *Personality: A Psychological Interpretation.* New York: Holt, Rinehart and Winston, 1937.

BEM, DARL, AND ANDREA ALLEN. "On Predicting Some of the People Some of the Time." *Psychological Review* 81 (1974): 506–20.

BOWERS, KENNETH. "Situationism in Psychology: An Analysis and A Critique." *Psychological Review* 80 (1973): 307–36

GOLDFRIED, MARVIN, AND RONALD KENT. "Traditional versus Behavioral Personality Assessment: A Comparison of Methodological and Theoretical Assumptions." *Psychological Bulletin* 77 (1972): 409–20.

KELLY, GEORGE. *The Psychology of Personal Constructs*, 2 vols. New York: W.W. Norton,Inc., 1955.

MISCHEL, WALTER. *Personality and Assessment*. New York: John Wiley and Sons, Inc., 1968.

———. "Toward a Cognitive Social Learning Reconceptualization of Personality." *Psychological Review* 80 (1973): 252–83.

MOOS, RUDOLPH. "Situational Analysis of a Therapeutic Community Milieu." *Journal of Abnormal Psychology* 73 (1968): 49–61.

four

General Comprehensive Texts

KAGAN, JEROME, AND H. MOSS. *Birth to Maturity*. New York: John Wiley and Sons, Inc., 1962.

MAIER, HENRY W. *Three Theories of Child Development*. New York: Harper & Row, 1969.

NASH, JOHN. *Developmental Psychology: A Psychobiological Approach*. Englewood Cliffs, N.J.: Prentice-Hall, Inc., 1970.

PARSONS, TALCOTT, AND ROBERT BALES. *Family, Socialization and the Interaction Process*. New York: Free Press, 1953.

SEARS, ROBERT, LUCY RACE, AND ROBERT ALPERT. *Identification and Child Rearing*. Stanford: Stanford University Press, 1965.

Ingestive Subsystem

JORDAN, H. A. "Direct Measurement of Food Intake in Man." *Psychosomatic Medicine* 28 (1966): 836–42.

LEVY, DAVID. *Behavioral Analysis*. Springfield, Ill.: Charles C. Thomas, 1958.

MANNING, MARY L. "The Psychodynamics of Dietetics." *Nursing Outlook* 13 (1965): 57–59.

RUBIN, RIEVA. "Food and Feeding." *Nursing Forum* 6 (1967): 195–205.

RUESCH, JURGEN. "A General Systems Theory Based on Human Communication." *General Systems Theory and Psychiatry*, ed. W. Gray, F.J. Duhl, and N.D. Rizzo. Boston: Little, Brown and Co., 1969.

STELLAR, ELIOT. "Hunger in Man: Comparative and Physiological Studies." *American Psychologist* 22 (1967): 105–17.

WALIKE, BARBARA, H. JORDAN, AND ELIOT STELLAR. "Studies of Eating Behavior." *Nursing Research* 18 (1969): 108–13.

WEINER, NORMAN. *Cybernetics*. New York: John Wiley and Sons, Inc., 1948.

Eliminative Subsystem

BEAKEL, NANCY, AND ALBERT MEHRABIAN. "Inconsistent Communications and Psychopathology." *Journal of Abnormal Psychology* 74 (1969): 126–30.

BUGENTAL, DAPHNE, JEROME KASWAN, AND LEONORE LOVE. "Perception of Contradictory Meanings Conveyed by Verbal and Nonverbal Channels." *Journal of Personality and Social Psychology* 16 (1970): 647–55.

DITTMAN, ALLEN T. *Interpersonal Messages of Emotions.* New York: Springer-Verlag, 1972.

EKMAN, PAUL, WALLACE V. FRIESEN, AND PHOEBE ELLSWORTH. *Emotion in the Human Face: Guidelines for Research and an Integration of Findings.* New York: Pergamon Press Inc., 1972.

FAST, JULIUS. *Body Language.* New York: M. Evans and Co., Inc., 1970.

KELLY, D., C. BROWN, AND J. SHAFFER. "A Comparison of Physiological and Psychological Measurements on Anxious Patients and Normal Controls." *Psychophysiology* 6 (1970): 429.

MEHRABIAN, ALBERT, AND M. WIENER. "Decoding of Inconsistent Communication." *Journal of Personality and Social Psychology* 6 (1967): 109–14.

OSGOOD, CHARLES, GEORGE SUCI, AND PERCY TANNENBAUM. *Measurement of Meaning.* Urbana, Ill.: University of Illinois Press, 1959.

RUESCH, JURGEN, AND W. KEES. *Nonverbal Communication: Notes on the Visual Perception of Human Relations.* Berkeley: University of California Press, 1961.

SCHLOSBERG, H. "A Scale for the Judgment of Facial Expressions." *Journal of Experimental Psychology* 29 (1941): 497–510.

STERNBACH, RICHARD. "Assessing Differential Automatic Patterns in Emotions." *Journal of Psychosomatic Research* 6 (1962): 87.

Sexual Subsystem

BARRY, H., M. BACON, AND IRVING CHILD. "A Cross-Cultural Survey of Some Sex Differences in Socialization." *Journal of Abnormal and Social Psychology* 55 (1957): 327–32.

BELLER, H. "Father Absence and the Personality Development of the Male Child." *Developmental Psychology* 2 (1970): 181.

BROVERMAN, I. et al. "Sex-role Stereotypes and Clinical Judgments of Mental Health." *Journal of Consulting and Clinical Psychology* 34 (1970): 1.

HARRIS, T. "To Know Why Men Do What They Do." *Psychology Today* 4 (1971): 35.

HOCH, P., AND J. ZUBIN, eds. *Psychosexual Development.* New York: Grune and Stratton Publishing Co., 1949.

LEVENTHAL, G. "Influence of Brothers and Sisters on Sex-Role Behavior." *Journal of Personality and Social Psychology* 16 (1970): 452.

MEAD, MARGARET. *Male and Female.* New York: Mentor Pub. 1955.

RESNICK, HARVEY. *Sexual Behaviors.* Boston: Little, Brown and Co., 1972.

ROSENBERG, G., AND BRIAN SUTTON-SMITH. *Sex and Identity.* New York: Holt, Rinehart and Winston, Inc., 1972.

SANTROCK, J. "Paternal Absence, Sex Typing and Identification." *Developmental Psychology* 2 (1970): 364.

SHILOH, AILON. *Studies in Human Sexual Behavior.* Springfield, Ill.: Charles C. Thomas, 1970.

STEPHENS, GWEN. "Mind-Body Continuum in Human Sexuality. *American Journal of Nursing* 70 (1970): 1468.

Affiliative Subsystem

BOWLBY, JOHN. *Attachment.* New York: Basic Books Inc., 1969.

DEMBER, W. "Birth Order and Need Affiliation." *Journal of Abnormal and Social Psychology* 68 (1964): 555.

GEWIRTZ, JACOB. *Attachment and Dependency.* Washington D.C.: Winston and Sons, 1972.

HARLOW, HARRY F. "The Nature of Love." *American Psychologist* 13 (1958): 673–85.

MACCOBY, ELEANOR, AND J. C. MASTERS. "Attachment and Affiliation." In *Carmichael's Manual of Child Psychology,* ed. Paul Mussen. New York: John Wiley and Sons, Inc., 1957.

ROBSON, K. "The Role of Eye to Eye Contact in Maternal-Infant Attachment." *Journal of Child Psychology and Psychiatry* 8 (1967): 13–25.

RUBIN, RIEVA. "Basic Maternal Behavior." *Nursing Outlook* 9 (1961): 683–86.

RUBIN, ZICK. "Lovers and Other Strangers: The Development of Intimacy in Encounters and Relationships." *American Scientist* 62 (1974): 182–90.

SCHUTZ, WILLIAM. *FIRO: A Three-Dimensional Theory of Interpersonal Behavior.* New York: Holt, Rinehart and Winston, 1960.

Dependency Subsystem

BERNARDIN, ALFRED, AND RICHARD JESSOR. "A Construct Validation of the EPPS with Respect to Dependency." *Journal of Consulting Psychologists* 21 (1957): 63–67.

CAIRNS, ROBERT B. "The Influence of Dependency Inhibition on the Effectiveness of Social Reinforcement."*Journal of Personality* 29 (1961): 466–88.

COX, F.N. "The Origins of the Dependency Drive." *Australian Journal of Psychology* 5 (1953): 64–75.

CUMMINGS, ELAINE, AND WILLIAM HENRY. *Growing Old: The Aging Process of Disengagment.* New York: Basic Books, Inc., 1961.

FITZGERALD, BERNARD. "Some Relationships Among Projective Tests, Interviews and Sociometric Measures of Dependent Behavior." *Journal of Abnormal and Social Psychology* 56 (1958): 199–203.

GEWIRTZ, JACOB. *Attachment and Dependency.* Washington, D.C.: Winston and Sons, Inc., 1972.

HEATHERS, GLEN. "Emotional Dependence and Independence in a Physical Threat Situation." *Child Development* 24 (1953): 169–79.

———. "Acquiring Dependency and Independence: A Theoretical Orientation." *Journal of Genetic Psychology* 87 (1955): 277–91.

JACUBCZAK, LEONARD, AND RICHARD WALTERS. "Suggestibility as Dependency Behavior." *Journal of Abnormal and Social Psychology* 59 (1959): 102–7.

KAGAN, JEROME, AND H. A. MOSS. The Stability of Passive and Dependent Behavior from Childhood through Adulthood." *Child Development* 31 (1960): 577–91.

KAGEN, JEROME, AND PAUL MUSSEN. "Dependency Themes on the TAT and Group Conformity." *Journal of Consulting Psychologists* 20 (1956): 29–32.

KALISH, RICHARD, ed. *The Dependencies of Old People.* Ann Arbor, Mich.: Institute of Gerontology: Wayne State University and University of Michigan, August 1969.

McCORD, W.J., AND P. VARDEN. "Familial and Behavioral Correlates of Dependence in Male Children." *Child Development* 33 (1962): 313–26.

Aggressive-Protective Subsystem

BANDURA, ALBERT. *Aggression: A Social Learning Analysis.* Englewood Cliffs, N.J.: Prentice-Hall, Inc., 1973.

BERKOWITZ, L. *Aggression: A Social Psychological Analysis.* New York: McGraw-Hill, Inc., 1962.

BUSS, ARNOLD. *The Psychology of Aggression.* New York: John Wiley and Sons, 1961.

CHOROST, S. "Parental Child-Rearing Attitudes and their Correlates in Adolescent Hostility." *Genetic Psychology Monographs* 66 (1962): 49–90.

DEUR, J., AND R. PARKE. "Effects of Inconsistent Punishment on Aggression in Children." *Developmental Psychology* 2 (1970): 403.

DOLLARD, JOHN et al. *Frustration and Aggression.* New Haven, Conn.: Yale University Press, 1939.

GARATTINI, P., ed. *Aggressive Behavior.* Proceedings of the International Symposium of the Biology of Aggressive Behavior. New York: John Wiley and Sons, Inc., 1969.

LORENZ, KONRAD. *On Aggression.* New York: Harcourt Brace Jovanovich, 1966.

SOLOMON, RICHARD. "Punishment." *American Psychologist* 19 (1964): 239–53.

Achievement Subsystem

ATKINSON, J.W., ed. *Motives in Fantasy, Action and Society.* New York: Van Nostrand, 1958.

ATKINSON, J.W., AND N.T. FEATHER. *A Theory of Achievement Motivation.* New York: John Wiley and Sons, Inc., 1966.

BREIT, S. "Arousal of Achievement Motivation with Casual Attributions." *Psychological Reports* 25 (1969): 539–42.

BRUCHMAN, I.R. "The Relationship Between Achievement Motivation and Sex, Age, Social Class, School Stream and Intelligence." *British Journal of Social and Clinical Psychology* 5 (1966): 211–20.

FRENCH, E. "Some Characteristics of the Achievement Motive in Women." *Journal of Abnormal and Social Psychology* 68 (1964): 119.

McCLELLAND, DAVID. *The Achieving Society.* New York: Van Nostrand Co., 1964.

McCLELLAND, DAVID et al. *The Achievement Motive.* New York: Appleton-Century-Crofts, 1953.

MOSS, H., AND JEROME KAGAN. "Stability of Achievement and Recognition-Seeking Behavior from Early Childhood through Adulthood." *Journal of Abnormal and Social Psychology* 63 (1961): 504–13.

MOULTON, R. "Effects of Success and Failure on Level of Aspiration as Related to Achievement Motives." *Journal of Personality and Social Psychology* 1 (1963): 399–406.

MUELLER, E. et al. "Psychosocial Correlates of Serum Urate Levels." *Psychological Bulletin* 73 (1970): 238.

OBERLANDER, M. et al. "Family Size, and Birth Order as Determinants of Scholastic Aptitude and Achievement in a Sample of Eighth Graders." *Journal of Consulting and Clinical Psychology* 34 (1970): 19.

ROTTER, JULIAN. "Generalized Expectancies for Internal versus External Control of Reinforcement." *Psychological Monographs* 80 (1966): 609.

WEINER, B., H. MACKHAUSEN, AND R. COOK. "Causal Ascriptions and Achievement Behavior." *Journal of Personality and Social Psychology* 21 (1972): 239–48.

Restorative Subsystem

HARTMANN, ERNEST. *Sleep and Dreaming.* International Psychiatry Clinics Series, vol. 7. Boston: Little, Brown and Co., 1970.

———. *The Functions of Sleep.* New Haven, Conn.: Yale University Press, 1973.

KLEITMAN, NATHANIEL. *Sleep and Wakefulness.* Chicago: University of Chicago Press, 1963.

OSWALD, IAN. *Sleeping and Waking.* New York: Elsevier Pub. Co., 1962.

five

BALDWIN, ALFRED, AND CLARA BALDWIN. "The Study of Mother-Child Interaction." *American Scientist* 61 (1973): 714–21.

BARKER, ROGER. *Ecological Psychology.* Stanford: Stanford University Press, 1968.

CANNON, WALTER. "The James-Lange Theory of Emotions: A Critical Examination and an Alternative Theory." *American Journal of Psychology* 39 (1929): 106.

ELKIND, DAVID, AND JOHN FLAVELL, ed. *Studies in Cognitive Development: Essays in Honor of Jean Piaget.* New York: Oxford University Press, 1969.

ENGEL, BERNARD, AND A. BICKFORD. "Response Specificity, Stimulus Response and Individual Response Specificity in Essential Hypertensives." *Archives of General Psychiatry* 5 (1961): 478.

FILBEY, R.A., AND MICHAEL GAZZINAGA. "Splitting the Normal Brain with Reaction Time." *Psychonomic Science* 17 (1969): 335–36.

FLAVELL, JOHN. *The Developmental Psychology of Jean Piaget.* New York: Van Nostrand Co., 1963.

GAZZINAGA, MICHAEL S. "One Brain—Two Minds." *American Scientist* 60 (1972): 311–17.

GOLDSTEIN, MICHAEL, AND J.N. ADAMS. "Coping Style and Behavioral Response to Stress." *Journal of Experimental Research and Personality* 2 (1967): 239.

GOLDSTEIN, MICHAEL et al. "Coping Style as a Factor in Psychophysiological Response to a Tension-Arousing Film." *Journal of Personality and Social Psychology* 1 (1965): 290.

GRIFFITT, W. "Environmental Effects on Interpersonal Affective Behavior: Ambient Effective Temperature and Attraction." *Journal of Personality and Social Psychology* 5 (1970): 240–44.

INSEL, PAUL, AND RUDOLPH MOOS. "Psychological Environments: Expanding the Scope of Human Ecology." *American Psychologist* 29 (1974): 179–88.

KIRITZ, STEWART, AND RUDOLPH MOOS. "Physiological Effects of Social Environments." *Psychosomatic Medicine* 36 (1974): 94–114.

KRECH, DAVID, AND RICHARD S. CRUTCHFIELD. *Elements of Psychology.* New. York: Alfred A. Knopf, Inc., 1962.

LACEY, JOHN I. "Principles of Automatic Response Stereotypy." *American Journal of Psychiatry* 71 (1958): 50.

———. "Psychophysiological Approaches to the Evaluation of Psychotherapeutic Process and Outcome." In *Research in Psychotherapy*, ed. E.A. Rubenstein and M.B. Parloff, pp. 106–208. Washington, D.C.: American Psychological Association, 1959.

LIBBY, WILLIAM L., JR., BEATRICE C. LACEY, AND JOHN I. LACEY. "Pupillary and Cardiac Activity during Visual Attention." *Psychophysiology* 10 (1973): 270–94.

LINDSLEY, DONALD B. "Attention, Consciousness, Sleep and Wakefulness." In vol. III. *Handbook of Psysiology*, pp. 1553–93. Washington, D.C.: American Physiological Society, 1960.

LURIA, ALEKSANDR K. *Higher Cortical Function in Man.* New York: Basic Books Inc., 1966.

MAGOUN, HORACE. *The Waking Brain.* Springfield, Ill.: Charles C Thomas, 1958.

MOOS, RUDOLPH. *Ward Atmosphere Scale Manual.* Palo Alto, Calif.: Consulting Psychologist Press, 1974.

————. "Conceptualization of Human Environments." *American Psychologist* 28 (1973): 652–65.

MURRAY, HENRY A. *Explorations in Personality*, Chapter 5. New York: Oxford University Press, 1938.

PETRIE, A. *Individuality in Pain and Suffering*. Chicago: University of Chicago Press, 1967.

PRIBRAM, KARL, AND DONALD BROADBENT, eds. *Biology of Memory*. New York: Academic Press, 1970.

ROESSLER, ROBERT. "Personality, Physiology and Performance." *Psychophysiology* 10 (1973): 315.

SCHACTER, S., AND JEROME E. SINGER. "Cognitive, Social and Physiological Determinants of Emotional States." *Psychological Review* 69 (1962): 379.

SCHULTZ, DUANE. *Sensory Restriction*. New York: Academic Press, 1965.

SOKOLOV, E.N. "Neuronal Models and the Orienting Reflex." In *The Central Nervous System and Behavior*, ed. Mary Brazier, pp. 187–271. New York: Joseph Macy, Jr. Foundation, 1960.

SOLOMON, PHILIP et al., eds. *Sensory Deprivation*. Cambridge, Mass.: Harvard University Press, 1965.

WENGER, MARION A. "Studies of Autonomic Balance: A Summary." *Psychophysiology* 2 (1966): 173.

six

ASCHOFF, J., M. FATRANSKÁ, AND H. GIEDKE. "Human Circadian Rhythms in Continuous Darkness: Entrainment by Social Cues." *Science* 171 (1971): 213–15.

BIRREN, JAMES E. "Translations in Gerontology—From Lab to Life; Psychophysiology and Speed of Response." *American Psychologist* 29 (1974): 808–15.

BIRREN, JAMES E. et al., eds. *Human Aging: A Biological and Behavioral Study*, Pub. No. (HSM) 71–9051. Washington D.C.: U.S. Government Printing Office, 1963.

BULLOUGH, BONNIE, AND VERN BULLOUGH. *Poverty, Ethnic Identity and Health Care*. New York: Appleton-Century-Crofts, 1972.

BURT, C., AND M. HOWARD. "The Multifactorial Theory of Inheritance and its Application to Intelligence." *British Journal of Statistical Psychology* 9 (1956): 95.

EYSENCK, HANS JURGEN. *Eysenck on Extraversion*. New York: John Wiley and Sons, Inc., 1973.

FESTINGER, LEON. *A Theory of Cognitive Dissonance*. Evanston, Ill.: Row, Peterson Inc., 1957.

FISHER, SEYMOUR, AND S. CLEVELAND. *Body Image and Personality*. New York: Van Nostrand Co., 1958.

GOLANN, S. "Psychological Study of Creativity." *Psychological Bulletin* 60 (1963): 548.

GRANICK, SAMUEL, AND R. PATTERSON. *Human Aging II: An Eleven Year Followup Biomedical and Behavioral Study.* U.S. Department of Health, Education and Welfare Publication No. (HSM) 71-9037, 1971.

HALL, CALVIN, AND GARDNER LINDSEY. *Theories of Personality.* New York: John Wiley and Sons, Inc., 1957.

HELLBRÜGGE, T. "Development of Circadian Rhythms in Infants." *Cold Spring Harbor Symposium on Quantitative Biology.* Vol. 25. Cold Spring Harbor, L.I., N.Y.: The Biological Laboratory, 1960.

HORNER, M. "Fail, Bright Women." *Psychology Today* 3 (1969): 36.

JACOBI, JOLANDE. *The Way of Individuation.* New York: Harcourt Brace Jovanovich, 1965.

KANGAS, J., AND K. BRADWAY. "Intelligence at Middle Age: A 39-Year Follow-Up." *Developmental Psychology* 5 (1971): 333–37.

MACCOBY, ELEANOR, AND CAROL N. JACKLIN. *The Psychology of Sex Differences.* Stanford: Stanford University Press, 1974.

NELSON R. "Childhood Interest Clusters Related to Creativity in Women." *Journal of Consulting Psychology* 29 (1965): 352.

NEUGARTEN, BERNIECE, AND RUTH KRAINES. "Menopausal Symptoms in Women of Various Ages." *Psychosomatic Medicine* 27 (1965): 266–73.

NEUGARTEN, BERNIECE et al. "Age Norms, Age Constraints and Adult Socialization." *American Journal of Sociology* 70 (1965): 710–17.

QUERY, WILLIAM T. *Illness, Work and Poverty.* San Francisco: Jossey-Bass Inc., 1968.

RICHTER, CHARLES P. *Biological Clocks in Medicine and Psychiatry.* Springfield, Ill.: Charles C, Thomas, 1965.

ROKEACH, MILTON. *Beliefs, Attitudes and Values.* San Francisco: Jossey-Bass Inc., 1969.

ROTTER, JULIAN. "The Role of the Psychological Situation in Determining the Direction of Human Behavior." *Nebraska Symposium on Motivation.* Lincoln, Nebraska: University of Nebraska Press, 1955.

SCHAIE, K. WARNER. "Translations in Gerontology—From Lab to Life. Intellectual Functioning." *American Psychologist* 29 (1974): 802–7.

SCHONFIELD, DAVID. "Translations in Gerontology—From Lab to Life. Utilizing Information." *American Psychologist.* 29 (1974): 796–801.

SEEMAN, MARVIN. "On the Meaning of Alienation." *American Sociological Review* 24 (1959): 783.

STEIN, MORRIS, AND SHIRLEY HEINZE. *Creativity and the Individual.* New York: Free Press, 1960.

SUEDFELD, PETER. *Attitude Change: The Competing Views.* Chicago: Aldine-Atherton Inc., 1971.

seven

BERNSTEIN, LEWIS, AND RICHARD DANA. *Interviewing and the Health Professions.* New York: Appleton-Century-Crofts, 1970.

CARLSON, CAROLYN, coordinator. *Behavioral Concepts and Nursing Intervention.* Philadelphia: J.B. Lippincott Co., 1970.

JACO, E. GARTLEY, ed. *Patients, Physicians and Illness,* 2nd ed. New York: Free Press, 1972.

KAHN, ROBERT, AND CHARLES CANNELL. *The Dynamics of Interviewing.* New York: John Wiley and Sons, Inc., 1957.

KAUFMAN, MARGARET. "High Level Wellness: A Pertinent Concept for the Health Professions." *Mental Hygiene* 47 (1963): 57–62.

LEBOW, MICHAEL. *Behavior Modification: A Significant Method in Nursing Practice.* Englewood Cliffs, N.J.: Prentice-Hall, Inc., 1973

MISCHEL, WALTER. *Personality and Assessment.* New York: John Wiley and Sons, Inc., 1968.

WU, RUTH. *Behavior and Illness.* Englewood Cliffs, N.J.: Prentice-Hall, Inc., 1973.

Index